Giulio Barbato

Alessandro Germak   Gianfranco Genta

Measurements for Decision Making

Measurements and Basic Statistics

Società Editrice ESCULAPIO

ISBN 978-88-7488-575-6

Società Editrice Esculapio s.r.l.
Via Terracini, 30 – 40131 Bologna
www.editrice-esculapio.com – info@editrice-esculapio.it

Editorial staff: Giancarla Panigali, Carlotta Lenzi

Translated by Anna Barbato

Società Editrice
ESCULAPIO

INDEX

1 INFORMATION CONTAINED IN MEASUREMENT RESULTS

1.1 INTRODUCTION

Measurement activities have well defined purposes. Generally people measure to know. In fact the empiric revolution given by Galileo Galilei brought to overtake ideal or even imaginative theories, convictions based on an inappropriate use of theology and *Ipse dixit*, giving the scientific knowledge more tangible and solid foundations.

But even the experimental data are not an absolute truth. It has to be evaluated and used within its limits in order to avoid misunderstanding.

This requires some considerations about the intrinsic characteristics of the information given by a measure[1] and to show how these characteristics are frequently inserted in international standard documents, used in the area of Quality. Knowing these premises, they will be easily comprehensible.

In the technologic use of measurement instruments it is evident how concepts like “exactness” or “equality” needs to be banned. That is hardly to do because the incorrect conviction that a result obtained with a high quality measurement instrument is exact is quite reassuring. If you can completely believe it, you will be also sure about the deduction that can be obtained. But exact is not a characteristic obtainable in real operations.

For instance if you have to design a hole-shaft junction, you could have the incorrect idea of producing it with exactly the dimensions required to obtain the amount of clearance necessary in order to allow the desired motion. However, since your initial mechanic technology experiences, you learnt that it is impossible. Therefore there is the need to operate in the scheme of ISO tolerance, which shows, more concretely, that the goal isn't the exact value, but a range called *tolerance*, that represents a reasonable agreement between functional needs and constructive possibilities. We must add that your hole-shaft connection can neither be made with exact dimensions, nor exactly measured.

With a simple practical exercise of repeating several times a measurement you generally obtain results containing differences, therefore you can get convinced that you don't ever get an exact result, but only a group of results different from each other.

It is clear now that this gives some problems highlighted by the following considerations:

- having obtained a group of different values, how can I deduce from them a single value that can represent the quantity measured?

[1] The term “measurement” is used to indicate the activity necessary to obtain a measure, which is the result of the measurement itself. Therefore we need to pay attention not to use the word measure with that meaning of measurement and vice versa.

- *realizing that the measurement result isn't exact, how can I evaluate the range in which it is reasonable to think that the value of the measured quantity is?*

The result dispersion that happens with no apparent rule indicates the presence of random factors and, therefore, the necessity to manage those problems with statistical methods. These will be useful even to give the answer, not sure, but reasonably probable, to another question that has to be always asked on the measurement results:

- is the variation of the results observed caused just by random factors or is there any specific cause producing it?

It is important to notice that this kind of question is present in many cases, even if phrased in a different way:

- are the workpieces, produced by two machine-tools, different just by chance, or because there is a real reason for the difference?
- are the measurement results suffering significantly because of variations of the environment conditions, or is it impossible to distinguish that from random effects?

It appears clearly that the same type of questions, specific cause or random effects, apply both in the field of measurement and production, underlining the opportunity of a statistical management of problems, necessary for treating random factors. Nowadays also design requires the application of statistics (*Statistical Tolerancing*).

1.2 THE INFORMATION CONTAINED IN MEASURES

It can be said that every measurement result is influenced by a multiplicity of factors, some of which are big and regular enough to allow the individuation of the relevant causes. These are *systematic* factors. Others, instead, are small, irregular, and cannot be described in a deterministic way; these are *random* factors, therefore need to be treated statistically.
The information contained in a measure can be only extracted taking into account that its representation through a numeric data has to be always considered distorted by systematic factors and confused by the random ones. It is important to consider always the interval of possible results, not only the value obtained in the measurement.

It is instructive to observe how the ideal meaning given to a measurement result can be misleading.

Examining a simple example of dimensional measurement, done with instruments of general use, we will see how deceptive the equality concept is and, therefore, also the fact to transmit the information contained in a measure just through the mere result.

In the shown case, the measurements of two gauge blocks[2] obtained using a micrometer (ISO 3611:2010) (Fig. 1.1), we can even see how, using only the measurement results, you can get a paradox.

In case of use of a modern micrometer with digital output, the measurement indication is given by a number that indicates the measured dimension with a reading resolution of 1 µm. If we measure two gauge blocks with the same nominal length with that micrometer, we should find almost certainly the same measurement value for both (the tolerance on the gauge blocks' length is generally less than 0.1 µm).

Should seem natural to say that ***their length is equal.***

But if we use an instrument with better sensitivity (VIM[3], 4.12), for example a Linear Variable Differential Transformer (LVDT) with a reading resolution of 0.1 µm or better, we should find, almost certainly, two different measurement results (Fig. 1.2).

We should conclude that their length is different!

It is evident that that would be clearly in contrast with what established before, with the micrometer.

The reason of the paradox is due to the false concept that says that the two blocks have a defined length value. Such concept, quite convenient to be used for defining ideal concepts, like the one of error, consist in believing that every physical quantity measured has a *true value*[4]. The simplifications connected to the notion of *true value* make it somehow considerable to the point of being included in the International vocabulary of basic and general terms in metrology (VIM, page viii and clause 2.11).

---

[2] Gauge blocks (ISO 3650:1998 Geometrical Product Specifications (GPS) -- Length standards -- Gauge blocks) are the mostly used length references in the industry. They are straight prisms made of metal (steel), synthetic (tungsten carbide) or even ceramic materials, characterized by having two opposite flat and parallel surfaces within really strict tolerances (few tenths of micrometre), which distance represents the reference length.

[3] JCGM 200:2012 (BIPM, IEC, IFCC, ISO, IUPAC, IUPAP, OIML) International vocabulary of metrology – Basic and general concepts and associated terms (VIM), 3rd edition, 2012, http://www.bipm.org/utils/common/documents/jcgm/JCGM_200_2012.pdf

[4] "true value of the physical quantity measured" is a pure abstraction, in terms of which it cannot be known (even taking several refined measurements, I will still have different results) or it doesn't even exist (It is enough to think about that, for example, in the length measures, exist instruments like Scanning Tunnelling Microscope (STM) or Atomic Force Microscope (AFM) that measure the molecular dimensions, therefore the length of a gauge block would depend on what molecules are taken as extremes. In such conditions it is almost impossible to define the measured entity). Despite this often people talk about "true value" of a measured physical quantity. For practical purposes, if the philosophical dilemma about the existence and usability of "true value" would be too pressing, it can still be used the concept of "reference value". This is the value of the measured entity, which can be concretely obtained using more refined instruments and systems, therefore measurable.

Operatively it is convenient to go from a comfortable, but illusory deterministic schematization to a more complex, but concrete statistical vision. We cannot think that a measure represents the value of the *measurand* (VIM 2.3), we can only identify a range of values around the result obtained, in which it is reasonable to think that it is contained the value of the measurand.

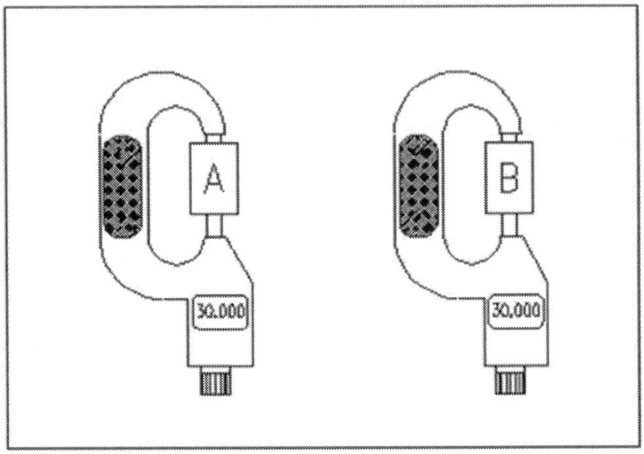

Fig. 1.1. The measurement, with a digital micrometer, of two gauge blocks having a nominal equal length, gives, generally an identical result.

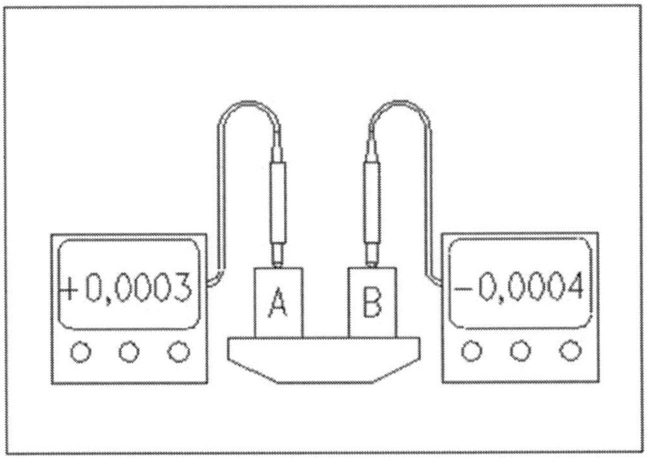

Fig.1.2. If the same blocks are measured with instruments having a higher resolution, it is quite possible to see that their dimension is different.

The consequence is that the information given by the measurement remains incomplete (Fig. 1.3) if to the result is not associated its uncertainty interval, which is a range around the measure obtained in which is reasonably believable that the reference value of the measurand is.

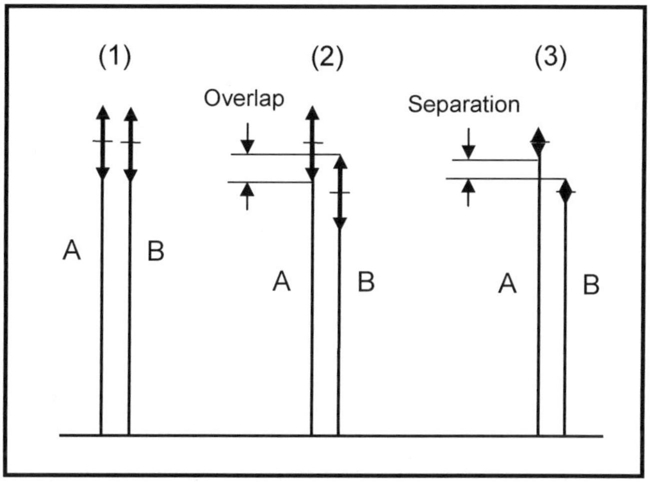

Fig. 1.3. The transfer of the information contained in the measures requires that the uncertainty interval is given. The equality described in (1) is only due to the large reading resolution and the information contained is not different from (2), where, being an overlap in the uncertainty intervals, it is impossible to tell which result is higher. In (3),instead, there is no overlap and it is possible to tell that the two results are different and that A>B.

We observe, in fact, the conclusions that can be obtained from the previous example.

In the reference case of the two gauge blocks, giving only the two measurement values, equal or different (i.e. A = 30.0003 mm, B = 29.9996 mm), doesn't allow to get conclusions about which block is longer. It is not excluded that another measurement would move the terms in the opposite direction. When the measurement values are completed by the extensions of the relevant uncertainly intervals, then it becomes possible to get from the data a coherent information (i.e. if an uncertainty of ± 0.4 µm is associated to the measurements you have to accept that the value given for A can be smaller, down to 29.9999 mm, and that the value given for B can be larger, up to 30.0000. There is a partial overlap of the uncertainty intervals as shown in Fig.1.3 (2), therefore it is impossible to know which one of the two blocks is longer. If instead to the same results an uncertainty of ±0.2 µm is associated, then the value of A cannot be less than 30.0001 mm and the value of B cannot be higher than 29.9998. Therefore there is a separation of the uncertainty intervals as shown in Fig.1.3 (3) and it is possible to affirm that A is longer than B, or that the two blocks are significantly different).

However it is never possible to establish an equality condition, therefore in case of overlap of the uncertainty intervals it is common to say that the two results are compatible. This indicates that their difference can be just an effect of random errors.

From this descends the objective importance of a correct evaluation of the uncertainty intervals, topic highly treated by the metrological organizations and international associations that have made the Guide to the expression of the uncertainty in measurement, usually called GUM[5], that was transformed in Europe and Italy in the ENV 13005, now withdrawn, and widely referred in ISO 14253-2[6].

To transfer the information produced by the measurement activity, it is then necessary to give a value and the relevant uncertainty.

1.2.1 Implications in Standard Specifications

The general and intrinsic conditions of the information contained in the measurement results have a repercussion on the operating rules and in Standard Specifications. We will talk about two different cases: a specific one, which exactly reflects the comparison above, and another one, more general, that implies how to manage the instrumentation for measurement and testing activities.

1.2.1.1 Guidelines on conformance or non-conformance with tolerances

The fundamental importance of the uncertainty evaluation is strongly underlined by the requirements of the ISO 14253-1[7] that says how to correctly respect tolerance. It is a comparison operation and conceptually it matches how said in the previous section. The only difference is that the comparison is often between a measurement result and a fixed value, not only between two measurement results.

In case of imposed tolerance by an external rule, like a law or a contract term that cannot be changed, it is important to take into account that the measurements taken during the test have their uncertainty. Therefore, even if the result is in the tolerance limit, it is possible that the measurand, that can have any value in the uncertainty interval, will be out of tolerance.

ISO 14253-1 requires the manufacturer to add to the lower tolerance limit an uncertainty half range and to remove the half range from the upper limit. The range of acceptance of the manufacturer is reduced compared to the tolerance range. Similarly, it has to be done by the customer, to evaluate a possible rejection, acting in the opposite direction, moving half uncertainty range under the lower limit and above the upper one, contracting the rejection

---

[5] JCGM 100:2008 (BIPM, IEC, IFCC, ISO, IUPAC, IUPAP, OIML), Evaluation of measurement data - Guide to the expression of uncertainty in measurement, http://www.bipm.org/utils/common/documents/jcgm/JCGM_100_2008_E.pdf

[6] ISO 14253-2:2011 Geometrical product specifications (GPS) - Inspection by measurement of workpieces and measuring equipment - Part 2: Guidance for the estimation of uncertainty in GPS measurement, in calibration of measuring equipment and in product verification

[7] ISO 14253-1:1998 Geometrical Product Specifications (GPS) Inspection by measurement of workpieces and measurement equipment. Part 1: Decision rules for proving conformance or non-conformance with specifications

area. As shown in fig. 1.4 instead of having neat borders in the tolerance range, there are, as limits, two uncertainty intervals

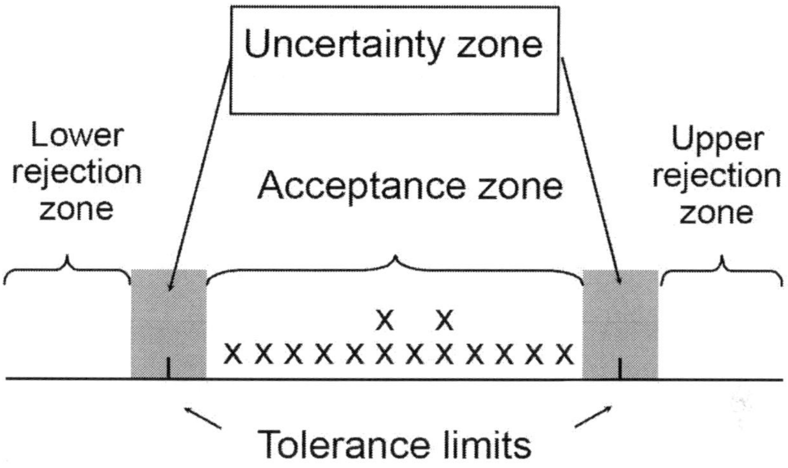

Fig. 1.4 The regulation ISO 14253-1 requires to define appropriate uncertainty zones in order to avoid, if possible, contrasts caused by the measurement uncertainty.

It is important to take note about such situation. Diminishing the acceptance range, the manufacturer increases the chance of rejection. At the same time for the customer, reducing the rejection range causes an increase of risk of receiving a piece out of tolerance and the impossibility of the customer to reject that product.

The advantage, however, is to diminish drastically the conflict possibilities between manufacturer and customer. The cost that any of them have to pay is as high as the uncertainty in the measurements, or as high as the probability to have a piece in the two uncertainty intervals. Here again the importance to consider the uncertainty and to do statistical considerations to decide correctly. This is really fundamental and avoids bad judgments. However, it is important to reduce the uncertainty in order to reduce costs and risks. So adopting the right metrological techniques and the correct uncertainty evaluation is the best way.

Always remember that the rules do not have to be taken as they are, but you need to understand the real reasons and advantages involved.

Let's talk about a concrete situation. In a well set factory, the general manager (Fig. 1.5) decided to have a small ordinary number of pieces out of tolerance, than to reach a zero-defects production (frequently very expensive for the changes required for the production plant and even not possible for high precision workpieces).

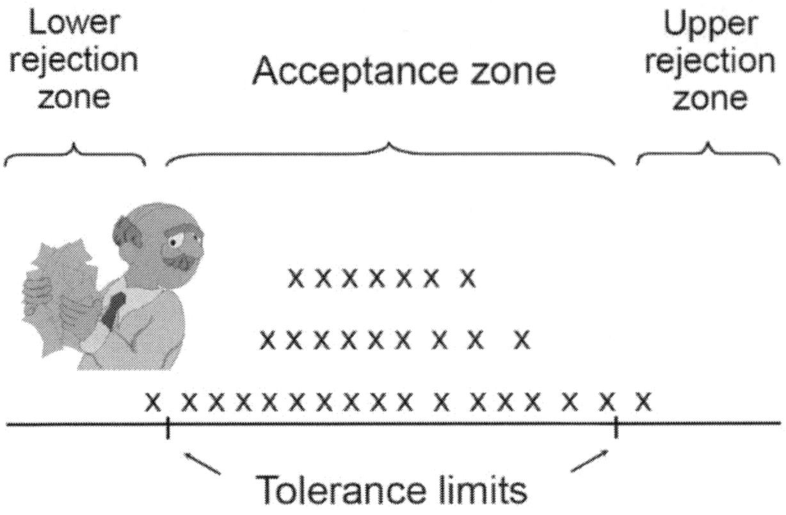

Fig. 1.5 The general manager is satisfied by a well set production with a limited number of rejected pieces.

The quality director follows an updating course, where he learns about the requirements of ISO 14253-1.

As soon as he arrives to the company, he immediately introduces the uncertainty intervals, with the direct consequence of doubling the rejected pieces.

Now the general manager is upset (Fig. 1.6). He has signed the Quality Policy that implies the obligation not to reduce the resources dedicated to quality, and even to increase them if it produces a higher customer satisfaction. Therefore he asked the quality director: "If until yesterday the quality level of a production with few rejected pieces was considered satisfactory, why, for the future, have we to accept to double the defect waste and the corresponding cost?". The quality director answered: "Because it is required by ISO 14253-1". An answer quite questionable!

Here we are again in front of a paradox. In this case, and in every other case where variations are proposed, often quite expensive, to a satisfactory situation, we need to deeply examine every element.

It is not possible to give an answer to the general director if the meaning of "tolerance" is not carefully considered. Tolerances can have at least three different reasons:

1. Imposed by law (for instance for safety reasons)
2. Agreed between producer and costumer
3. Required for technological reasons

When tolerances are imposed by law, it is fundamental to satisfy the ISO 14253-1 requirements, to be sure that the uncertainty in your test does not allow a product out of tolerance to pass your quality control.

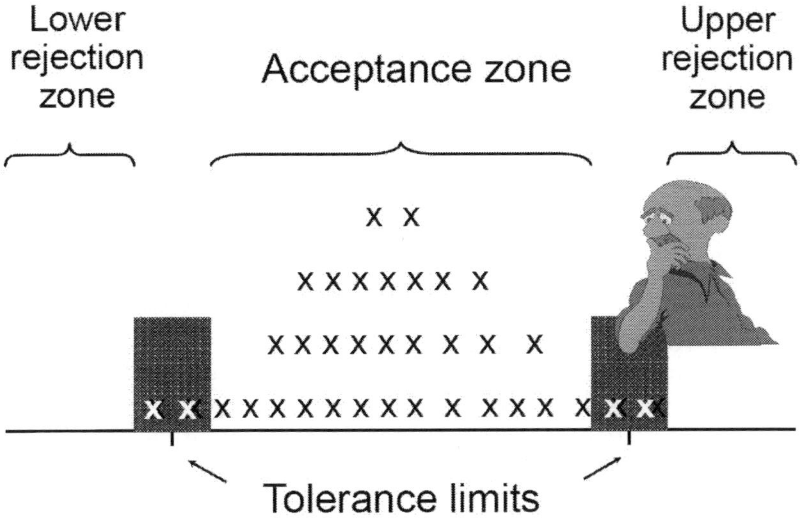

Fig. 1.6 The quality director gets the information about ISO 14253-1 and he introduces the uncertainty intervals. The acceptance area is reduced and the defect waste doubles. Surely the general manager is not happy about this situation!

When tolerances are agreed in a producer-customer contract, possible changes shall undergo the same way and agreed between producer and customer.

Most frequently tolerances are due to technological reasons: in a production process testing actions, with relevant tolerances, are inserted to avoid that a workpiece is passed to a successive production step if inadequate.

For instance, before of a grinding operation, the workpiece hardness shall be controlled. In fact grinding is not successful on materials not properly hardened, even the grinding wheels become clogged with the soft material, causing serious damage to the production process. Therefore a hardness test is required and a tolerance like 52 ± 2 HRC is established. It is important to underline that such limit is not imposed by a law or other mandatory rule. It is established by the manufacturer through a series of tests finalized to evaluate what happens in the following processing phase, with a variation of hardness.

The tolerance limits position, for example of the lower limit of hardness value, as represented in fig. 1.7, is related to the positive or negative behaviour of the piece in the following working phase and to its functionality and reliability.

For too low results in hardness the piece will always be unsuitable (two "r" full columns).

For intermediate results it can be sometimes unsuitable, sometimes acceptable (third column with "r" and "a").

Finally, when the results will be adequately high, the sequent behaviour of the piece will always be good (columns with all "a"). Consequently the tolerance limit is established.

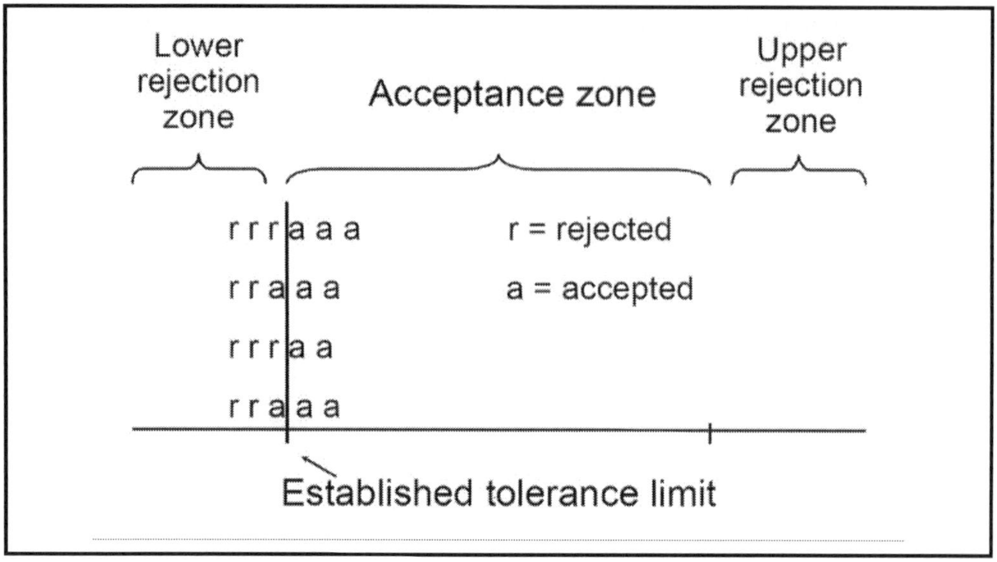

Fig. 1.7 When the tolerances aren't imposed by legal or contractual obligations, they get determined with tests. These are made by varying the examined parameter and determining the success of the produced piece (accepted, ***a****) or its discard (rejected,* ***r****). Such determined boundary takes into account of the uncertainty of the used instruments; therefore it doesn't need to be moved.*

Summarizing in general, considering the variation of test results on an examined characteristic, it will be possible to see that:

- Below a certain value, the behaviour of the piece will always be unsuitable;
- In case of intermediate values, the results may vary. There will be negative or acceptable conditions;
- At last, over a certain value there will always be acceptable conditions.

The latter value is the one that is chosen, generally, as lower limit of the tolerance. It is important to notice that in this limit selection process the measurement instrument uncertainty is involved. Then the uncertainty zone coincides with the second case listed before. It is evident that, if the limit has been determined in the described way, there is no reason to move it more, reducing, in an unjustified and useless way, the acceptance area.

But the uncertainty of the instruments used during the tests has to be similar to the ones used to set the limit itself[8].

[8] Sometimes, setting a production process, testing instruments are better than the ones used in normal production control. In this case evaluation of the uncertainty of the two types of instrument shall be done and resulting adjustments applied following ISO 14253-1.

1.2.1.2 The management of measurement and test activities according to Quality rules

The majority of the information that define and qualify products are obtained with measurement operations. Therefore the quality standards contain specific indications about the management of measurement activities.

Particularly, ISO 9001:2008[9] requires that the measurement processes are adjusted to the qualification needs of the product or trial. A substantial point regarding measurement results is their general compatibility[10]. The purpose is to make sure that there won't be any protest due to the malfunctioning of the instruments. It is better to underline as soon as possible that such condition, even if necessary, is not sufficient to grant the result compatibility. In facts, as we will see, the measurement result depends on the whole *measurement complex*[11]: instrument, operator, environmental conditions and the measurand[12].

In case the product is a measure (testing rooms, metrological rooms, calibration services), its characteristics are completely defined. In facts the quality of any measure can be evaluated through the uncertainty related to it. This is the reason why there is a strong difference between the general quality rules and the specific ones related to the measurements. The latter can be applied to calibration and test laboratories. Such difference is underlined by the terminology. We talk about *certification* in quality of a factory and about *accreditation* of a calibration or test laboratory. The difference is highly important. The certifying Authority cannot control the ability of a certified factory to make an acceptable product, because the acceptability criteria are various and owned by the certified factory and not by the Authority. The accreditation Authority should be able, directly or indirectly, to do the measurement operation better than the accredited laboratory. Therefore it is able to evaluate the goodness of the results obtained by the laboratory.

The two conditions of certification and accreditation follows different kind of Standard documents. In the certification area the principal Standard document is ISO 9001:2008. In the accreditation area the standard document is ISO 17025:2005[13].

---

[9] ISO 9001:2008 Quality management systems -- Requirements

[10] Ideally you can think that measurement results from different labs should be the same, but it never happens. What is important is that the compatibility, as described in 1.2 and in Fig. 1.3(2), of the data produced by different instruments must be proven (traceability, VIM 2.41).

[11] ISO/IEC 17025:2005, General requirements for the competence of testing and calibration laboratories.

[12] If you think, for instance, that a ball bearing can be idealized with a geometric sphere; it would be possible to qualify it by a single diameter, because in a geometric sphere every diameter is equal. In the real piece diameters are different (measurement is frequently done with 0.1 μm resolution, having differences of few micrometres). Therefore in the measurement process the real condition of the measurand produces effect on the measurement results and on their uncertainty.

[13] ISO/IEC 17025:2005, General requirements for the competence of testing and calibration laboratories.

While in the old way the system conditions were mandatory, in the new one any action required is always subordinated to the necessity imposed by functional specifications[14]. This is quite important and strongly changes the action line, moving it from a formal agreement that often didn't match high costs with real advantages, to a way of individuation and solving problems. But this line of action has as prerequisites that involved technicians have a deep knowledge on the specific matter.

As usual there is not an optimum solution valid for all the situations. Sometimes, the automatic formal answer, even if oversized to the real needs, is, however, the most advantageous because it doesn't require a complex analysis of different problems. In other cases, the formal answer's cost would be clearly higher than a targeted response to the real needs. It is important to check the alternatives being able to defend our own choice.

For the purposes of this chapter, aimed especially at understanding how important is the uncertainty of measurement for all activities, it is enough to say that test and measurement uncertainty evaluation is explicitly required by Quality rules. At managerial level, it is one of the pillars in order to pass from measures to decisions.

Frequently we heard someone saying: "We worked for years with our and our clients' satisfaction, without even knowing what the measurement uncertainty was. Why should we now spend time and money to evaluate it?".

What to answer? Certainly not that one has to do it, because it is required by Technical Standards. In fact, Quality Standards requires it "... when necessary".

First of all we have to say that it is not true that customer satisfaction was obtained for years without even knowing the concept of uncertainty. Tolerances, that are widely used, represent the translation of workpiece functional specifications and, as seen before with the effect of uncertainty on conformity assessment, the indication of acceptable uncertainty and the way for deciding which testing instrument is suitable. The selection of measurement instruments and methods was always made by competent people, focusing on the production objectives and on the "measurement capability" of the instruments set by their own judgement, which is practically the memory of previous correct decisions but also of previous errors.

A great benefit given by the application of the Guide to the expression of the uncertainty in measurement (GUM) is in the help for avoiding the cost of building the experience also by errors. GUM gives a practical way for getting the maximum advantage from the use of every previous technical experience, managed in a controlled theoretical and practical scheme, for getting the expression of uncertainty.

[14] ISO 17450-2:2012 Geometrical product specifications (GPS) -- General concepts -- Part 2: Basic tenets, specifications, operators, uncertainties and ambiguities

2 QUALITY MANAGEMENT OF MEASUREMENT AND TESTING

2.1 INTRODUCTION

In the previous chapter, the main topics have been:

1) the real information contained in measurement results[1];
2) some general and specific rules, pertaining to the management in a Quality System of measurement and testing activities.

Both have limits and advantages, and point out to the importance of organizing any action basing it on real needs compared with the relevant uncertainty.

This way is clearly depicted by ISO 17450-2[2] that describes how to face any production or measurement problem (Fig. 2.1). It confirms that the main characteristic for evaluating the Quality of a production or measurement process is the uncertainty of the product (workpiece for production, measure for measurement). Moreover it stresses that the starting points are the real needs of the process. This can be described by an example due to Johan Dovmark: design and verification of a dispenser (Fig. 2.2).

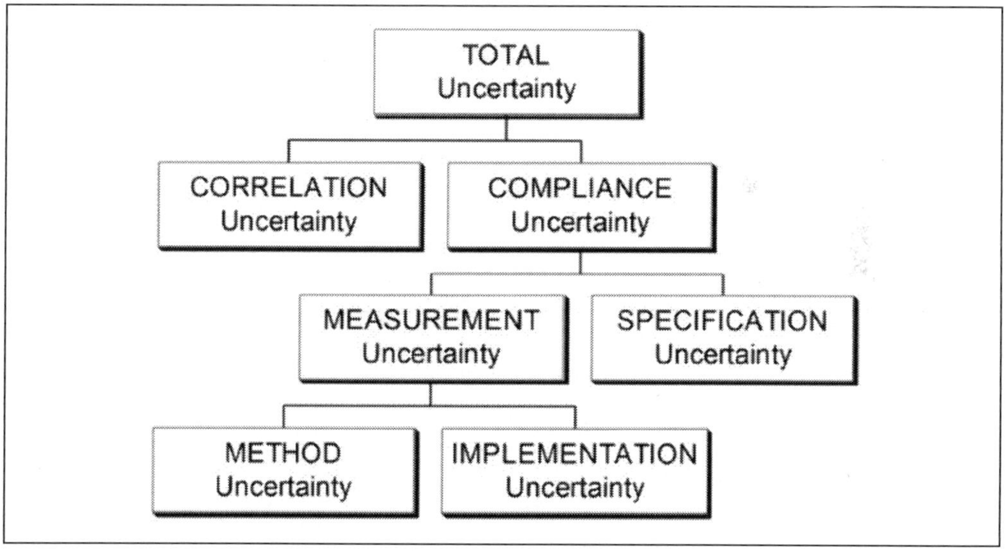

Fig. 2.1 Uncertainty scheme proposed by ISO 17450-2

---

[1] It's the conceptual base that we have to keep in mind to avoid paradoxes, otherwise unexplainable, and to understand the reasons of some law requirements.

[2] ISO 17450-2:2012 -- Geometrical product specifications (GPS) -- General concepts -- Part 2: Basic tenets, specifications, operators, uncertainties and ambiguities

The dispenser is a cylinder containing the material to be dispensed that shall be hand-operated: acting with the thumb, a piston is moved and the material out of the cylinder is dispensed.

Point 1. (Total Uncertainty). We must have a clear understanding of the functional specifications, which are the working conditions of the device we are designing. For clearness we focus only on the mechanical specification: we want that the force necessary for operating the dispenser be so low to allow hand operation, but not too low to risk unwanted dispensing. This can be subdivided in two parts: establishing a correlation connecting functional specification with measurable characteristics and checking the compliance of the produced workpiece with those characteristics.

Point 2.1 (Correlation Uncertainty). Translation in operational specifications. A first simple translation of these conditions in mechanical terms is that the operating force shall be between 5 N and 20 N. This condition is obtained by means of two circular rings (one on the cylinder, one on the piston) that define a stopping state which is eliminated because they break under a force. The second, more complicated, translation of functional specification shall indicate the dimension of the retaining rings that allow the breaking in the required range of force. This condition shall be evaluated by a specific experiment. Let's consider all the dimensions fixed except for the hole diameter (the ring on the cylinder), whose change produces a variation of the breaking force. Therefore, the experiment shall correlate the force amount with diameter value. In our case, the experiment showed that the diameter between 8.0 mm and 8.1 mm is suitable for the required force range (Fig. 2.2).

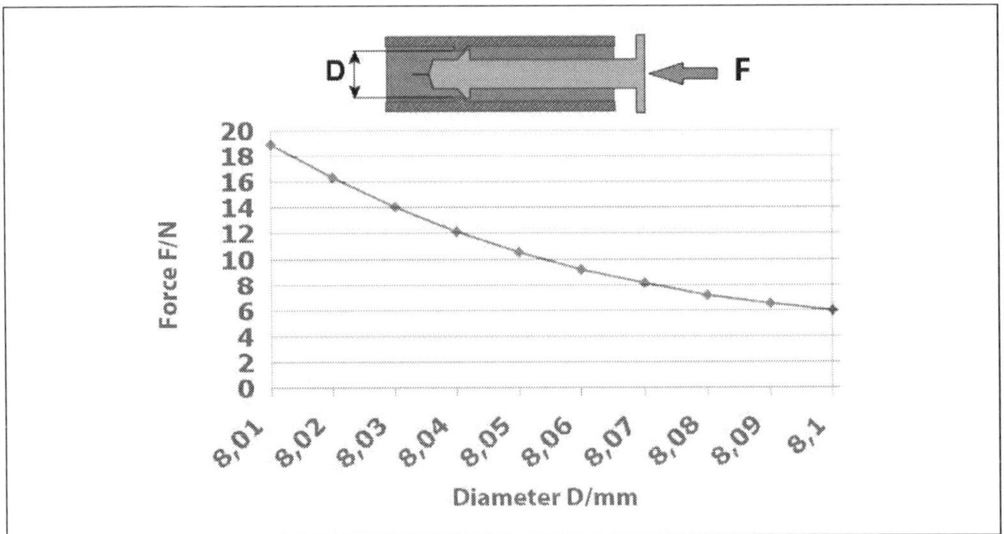

Fig. 2.2 The dispenser and its Correlation Characteristic of the breaking conditions (obtained experimentally). The uncertainty on the definition of the correlation characteristic and on the relevant experimental results is called correlation uncertainty

As for any evaluation, this cannot be done exactly, therefore we shall consider the correlation uncertainty as a contribution to the total uncertainty.

Point 2.2 (Compliance Uncertainty). Consists in evaluating the compliance of our workpiece with the specifications as translated before in dimensional requirements. This can be subdivided in two independent parts: 2.2.1) describes thoroughly the geometrical requirements with geometrical specifications, and 2.2.2) checks with measurement the conformity of the workpiece to the given specifications.

Point 2.2.1. (Specification Uncertainty).To inform the workshop about the dimension and tolerances of the workpiece we must give the dimensional specifications: values and tolerances. In our case we shall transmit to the workshop the information about the diameter. The problem is that the diameter is not a geometrical characteristic univocally defined. In fact, the real ring has certainly a circularity error (the diameter is not constant); therefore we shall consider which type of diameter is important for the breaking conditions. Usually three types of diameter can be defined. An outer one, outside the object (note that for a hole this is the minimum diameter, while for a shaft this is a maximum diameter), usually called maximum material diameter, an inner one, touching the more internal points of the object (for a hole this is the maximum diameter, for a shaft is the minimum diameter), called minimum material diameter. Eventually, we can be interested in a medium circle, balancing the more external with the more internal parts, usually obtained with a least squares (LS) evaluation (Fig. 2.3).

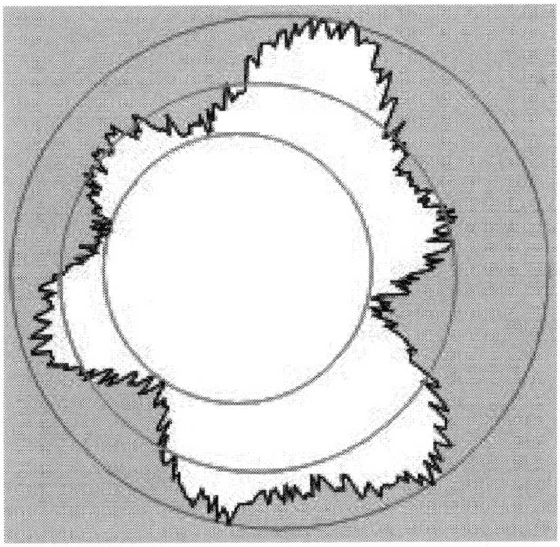

Fig. 2.3 Considering the real irregular shape of a hole, three circles can usually be defined: one inside the material, called "Minimum material circle" (the larger one), one outside the material, called "Maximum material circle" (the smaller one), and eventually a circle balancing the irregularities using the least squares method, called "Least squares circle"

Specifications shall be given correctly. If they are not completely defined (e.g. frequently is given only a dimensional tolerance on diameter, not associated with the condition of maximum material, minimum material or LS evaluation) or wrongly defined, we have for this cause a contribution of uncertainty, called specification uncertainty.

Point 2.2.2 (Measurement Uncertainty). It consists in evaluating the conformity of our workpiece to the given specifications, usually doing a test with some measurements. We can, therefore, consider measurement uncertainties.

We can subdivide this in two independent parts: 2.2.2.1) the test method adopted, and 2.2.2.2) the way the test is implemented.

Point 2.2.2.1 (Method Uncertainty). For instance, we can decide to find the diameter identifying a circle by three measurement points (wrong decision!). Even worst is, for allowing an easy measurement procedure, to situate the three points on a small part of the circle. In fact, in case of shape errors (always present) the measured diameter changes if the measurement points are changed, with a really large variation if we take only three points close together (Fig.2.4). If we choose to identify the circle with a large number of points, well-disposed all around, the risk to get significantly different diameter values using different sets of measurement points is small. The amount of possible changes involved is called Method Uncertainty.

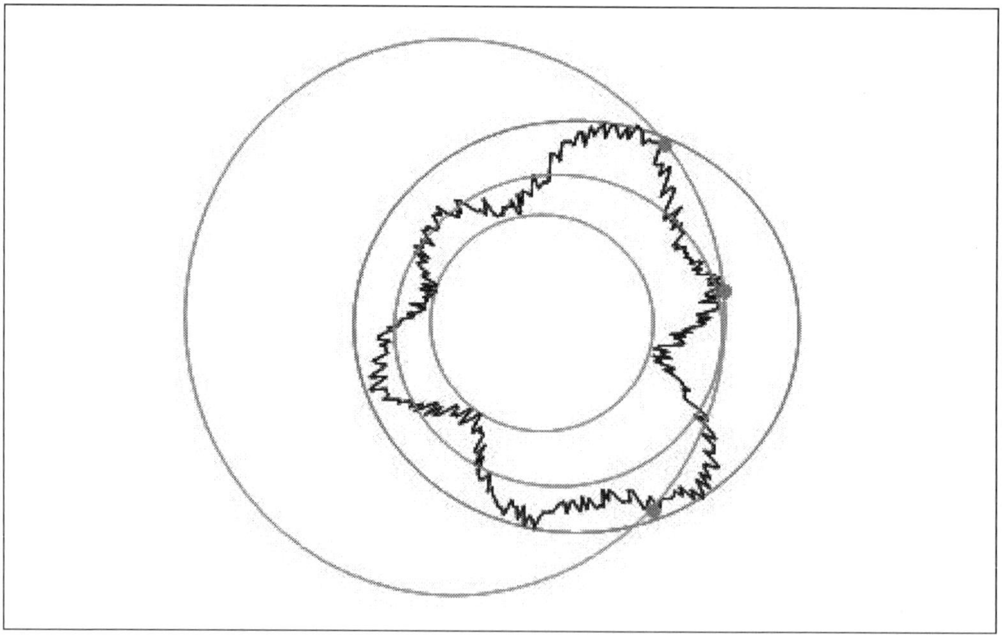

Fig. 2.4 Considering the real irregular shape of a hole, if we try to identify a circle using three measurement points close to each other, we can get very erroneous results.

Point 2.2.2.2 (Implementation Uncertainty). Using the same method we can get different measuring variations depending on the way we perform our test. We can use instruments with outstanding or normal metrological characteristics. Our operator can be expert or just informed about the procedure. The environmental conditions can be well controlled, like in a metrological lab, or highly variable. The measurand can be well defined, a nearly perfect circle, or highly irregular. All this conditions produce variations of results, related to the Implementation uncertainty.

The scheme described before can give a great help for avoiding and solving problems, but frequently is not considered in the industrial practice. Two real examples can be given with respect to tolerancing.

2.1.1 The best possible (sometime impossible) realization

During a conference at an industrial site I was asked by a designer to explain to his manager that he was making an impossible request. The problem was that the usual mechanical production had dimensions in the range of 200 mm, while a new request involved dimensions ten times larger. The position of the manager was: "... our usual tolerance of 0.01 mm shall be applied". It is very common that tolerancing do not follow the correct decision process:

functional specifications → production requirements

but they are just established conventionally ("We will do as we have always done!") or, even worse, they are defined considering the best possible working conditions. In this case, the situation was solved because the manager required almost impossible tolerances. In other cases the required tolerances were possible, even if really expensive, and applied. The problem will be showed up later.

I was asked by a producer of mechanical devices: he complained because the Quality Inspectors did not accept his measurement instruments. I knew the functional requirements for his production and that the tolerance should be around 0.03 mm, therefore the instrument used were perfectly suitable. During the discussion the producer said that he knew that a tolerance of 0.03 mm was sufficient, but to be sure that the workshop should do its best, he put in his blueprints tolerances of 0.01 mm. Analysing the situation, it was clear that the Quality Inspectors were right. A useless tolerance produced a significant damage.

A further example can be given about an industrial activity carried out as a stage by one of my students. A piece of car body (strut) had to be checked. It is a piece inside the engine hood, between the passenger compartment and the front side. Its task is to correctly position the fender and the front side with respect to the main frame. It was considered as a part of the coachwork, therefore its usual tolerances were applied: 0.5 mm all over the surface. The verification gave a lot of points out of tolerance (fig. 2.5). As frequently happens, the measurement laboratory was considered guilty, and new measurements, done by a different laboratory were required. But, as shown in the Fig. 2.5, the results where compatible and most of the points confirmed out of tolerance. They had to decide what to do and my student reported that, after some days, a young engineer (the old team leaders were absolutely sure about the correctness of the traditional procedure) questioned about the real task of the strut. Not the aesthetic part, as the exterior side of the coachwork, but the correct distance between some points of the main frame and the supporting points of the fender and the front.

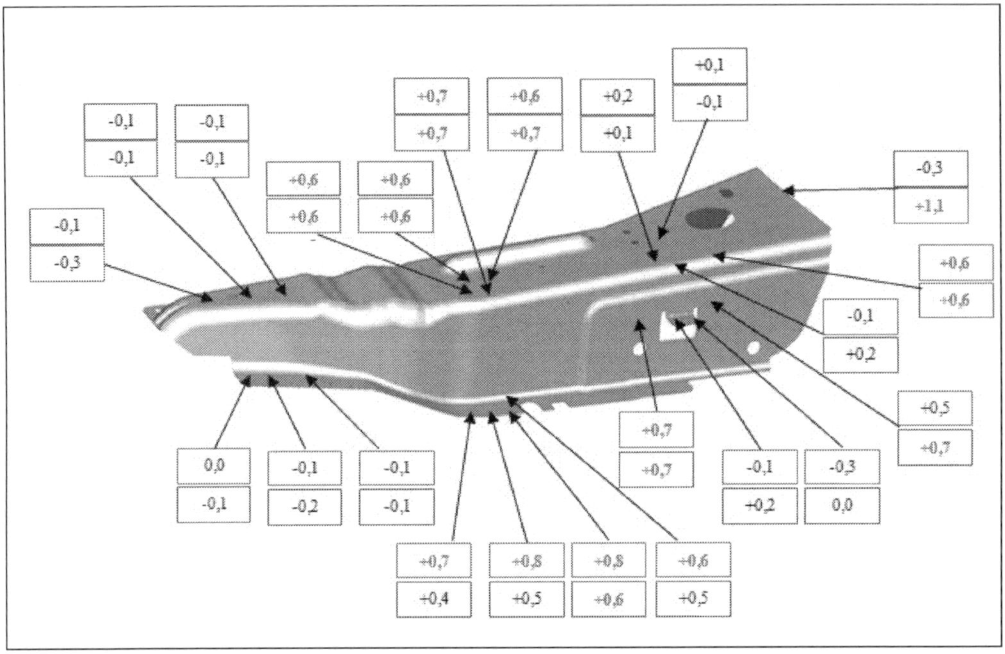

Fig. 2.5 Verification of a car workpiece. Measurements taken in two different laboratories confirmed that many points were out of tolerance.

Specifications were drastically reduced, as shown in fig. 2.6. This shows how easy is to waste days of work if the functional specifications are not considered deeply from the beginning. But as we can see in table 2.1 some out of tolerance remains. It was necessary to analyse data for understanding if there was something systematic. Statistical data analysis allows to evidence if there are systematic effects and what causes them.

This is a very important help for obtaining a better production with minimum expenses. In fact, systematic effects can be corrected, compensated or the cause producing them eliminated with specific actions.

This is definitely less expensive than a general operation necessary to face random effects. Therefore, the statistical methods, allow to understand if there are systematic effects, to connect them to their cause, and to make a mathematical model of the cause-effect relationship results very useful.

We will deeply see that after, but now we see here just a tip of experience. The concepts systematic and random are strongly connected with the visible condition of order and disorder. Therefore, if we can see a certain type of order conditions in our data, we get warned that there could be something systematic.

Looking at table 2.1 it seems evident that there is a clear order condition: the points out of tolerance are all positive. This gives us the indication of a displacement of the workpiece in a given direction. The piece was posed in a wrong position on the Coordinate Measuring Machine (CMM) and the control was made on the point coordinate (position) not on their reciprocal distance and angles (shape). To correct this we used the roto-translation mathematical tool, based on the statistical method of regression, and we got the results reported in table 2.1.

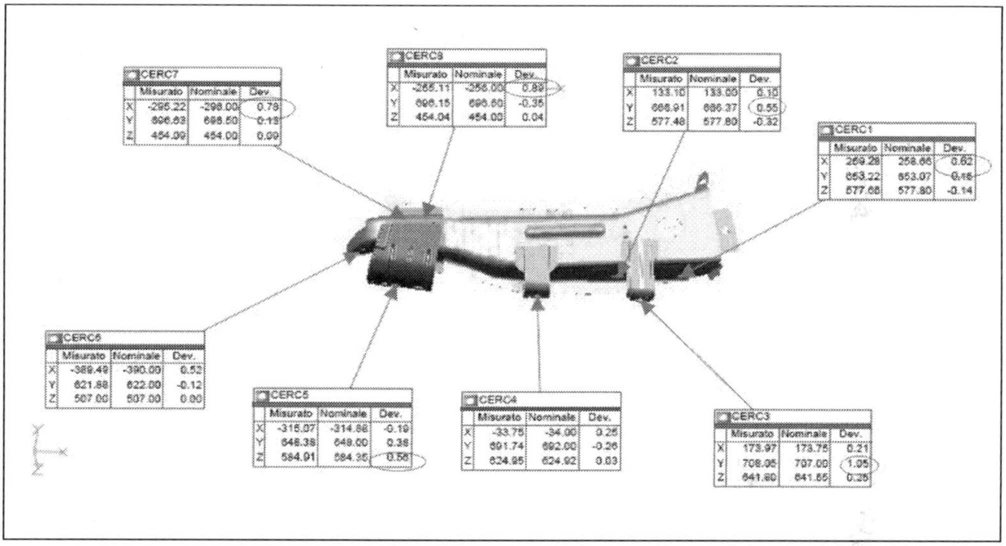

Fig. 2.6 Considering the real functional specifications of the workpiece, only few points shall be controlled. The number of out of tolerance is drastically reduced, but some remains.

The mathematical correction of the workpiece wrong position has produced its effect: all the points considered out of tolerance are, now, acceptable, while the real coordinate out of tolerance, Δy on CERC4, was not evidenced by original data.

In this case the specification uncertainty applies because, instead of using the necessary shape tolerances, coordinate tolerances were given, or used. Moreover it shows how useful can be a data analysis to point out the causes that produce unacceptable effects.

The discussion given hereby indicates that the problems can be quite different and that the application of general rules of good practice, like the protocol of ISO 17450-2, requires to think deeply and carefully about the problem conditions and to use a specific experience.

Tab. 2.1 Coordinates and errors from the original data and after roto-translation. The number of out of tolerance (tolerance is 0.5 mm) is reduced from 6 to 1. Note that the real out of tolerance point (Δy on CERC4) was considered as acceptable before the position correction.

Point	Original data						Roto-translated data					
	x	Δx	y	Δy	z	Δz	X	Δx	y	Δy	z	Δz
CERC1	258.66	**0.62**	653.07	0.15	577.80	-0.14	258.68	-0.20	653.61	0.39	577.55	-0.11
CERC2	133.00	0.10	666.37	**0.55**	577.80	-0.32	132.99	0.29	666.80	-0.12	25.01	0.22
CERC3	173.75	0.21	707.00	**1.05**	641.65	0.25	173.57	0.01	707.58	-0.47	89.02	-0.12
CERC4	-34.00	0.25	692.00	-0.26	624.92	0.03	-34.28	-0.13	692.36	**0.62**	72.41	0.22
CERC5	-314.88	-0.19	648.00	0.38	584.35	**0.58**	-315.28	0.19	648.02	-0.36	31.91	-0.27
CERC6	-390.00	**0.52**	622.00	-0.12	507.00	0.00	-390.13	-0.25	621.83	-0.05	-45.64	0.12
CERC7	-296.00	**0.78**	696.50	0.13	454.00	0.09	-295.52	0.10	696.37	-0.26	-98.70	-0.04
CERC8	-256.00	**0.89**	696.50	-0.35	454.00	0.04	-255.50	0.01	696.41	0.26	-98.74	-0.02

Some matter on the measurement activity can be established by more specific and practically applicable rules as described in the Quality Standards pertaining to measurement and testing management. However, industrial applications involve such differences that oblige to pay attention and to understand the reasons of any action, allowing routine only in few cases. This can give a good light on this concept:

> *JCGM 100:2008*[3] ***3.4.8*** *Although this Guide provides a framework for assessing uncertainty, it cannot substitute for critical thinking, intellectual honesty and professional skill. The evaluation of uncertainty is neither a routine task nor a purely mathematical one; it depends on detailed knowledge of the nature of the measurand and of the measurement. The quality and utility of the uncertainty quoted for the result of a measurement therefore ultimately depend on the understanding, critical analysis, and integrity of those who contribute to the assignment of its value.*

"Critical thinking", "Intellectual honesty", "Professional skill". Three characteristics necessary when we shall go the path described before in the uncertainty scheme.

Standards can give a general, frequently only generic "Framework" as a guide. Quality Standards acts in the same way, but it commonly happens that are taken as an absolute law, despite the great attention of the experts which have written the standards to refers to what is necessary, the functional specifications, and not to give an absolute and rigid line of action.

[3] BIPM JCGM 100:2008 document Evaluation of measurement data – Guide to the expression of uncertainty in measurement in: http://www.bipm.org/en/publications/guides/gum.html

To understand this is sufficient to consider the main different industrial production conditions, in reference to what seems to be a formal Quality rule asking for the calibration of each measurement instrument. Let's see the implications of this in a common industrial process: a continuous process, characterized by the only necessity of keeping constant in time the most important parameters.

An example of continuous process is the industrial heat treatment. Setting the process at the beginning makes possible to find experimentally a good temperature-time conditions. Thereafter the product quality is obtained just keeping those process parameters at the initial values. The metrological characteristic of the measurement instrument used to verify the working conditions is, therefore, only the stability. This is the ability of a measuring instrument to give nearly constant results for the time required. We are not interested to have a given value of the temperatures (this require calibration of thermometers), but only that the temperatures remain adequately constant until the process continue.

Different is the condition when the producer guarantees the customer a given characteristic of his product. For example, if we take as a product a cylindrical pin with the maximum material diameter of 10 mm and a cylindricity tolerance of 0,01 mm. In this case the producer shall have a calibrated instrument and a testing process with an uncertainty in the range of 2 µm. This is certainly expensive, because, at that uncertainty level, it should be done in a temperature controlled environment like a metrology room. Generally this is practically impossible, because it is really difficult to check all pins in a metrology room.

The best way is to switch from measurement to comparison. We can characterize some pins, carefully produced and identified by a serial number, in the metrology room at a temperature near to the 20 °C reference temperature. Then we take some pins with a more regular shape and, if possible, but not necessarily, close to the nominal value of diameter. For instance, the pin s/N 0001 is characterized by a maximum material diameter of 10.006 mm and a cylindricity error of 0.003 mm. If we refine our measurand, that is not any diameter of our pin, but a well-defined and identified diameter (in a given section and direction), we can reduce the variability. We can choose a diameter of 10.004 mm which measurement can be repeated within 0.001 mm, which is now our reference diameter. Now we can organize in line our comparison test: we will not find the value of the diameter of all the pins produced, but we compare them with our reference diameter. Therefore, in line, we can use a comparator that is usually more simple, rugged and easy to use than a measurement instrument. We set it to 0.004 mm measuring few times our reference diameter, taking the average and adjusting our comparator to the average. Now, if we check a pin on a diameter of 9.998 mm, we will get from the comparator the result -0.002 mm. In this way, we can check our production by comparison, which involves only the precision of our comparator, which can be checked repeating measurement on our reference diameter; therefore no calibration is required. Note that, as you take our reference pin from our production, made of the same material of the pins you will check, the temperature effect is automatically compensated. If in summer in our production workshop we have a temperature of 30 °C, our 9.998 mm pin increases of about 1 μm. But also our reference pin is increased of 1 μm. Therefore the zero of our comparator is now 0.001 mm and the difference remain 2 μm as before.

We can understand, from these examples, that the calibration is not always necessary. Calibration and environmental conditions associated to a complete measurement process are very expensive, therefore any time we are able to switch from a complete measurement to a simpler check of stability or of comparison we save a lot of money.

If we read carefully the main Quality Standards, we will see that even there the reference to the functional specifications is considered important: "Only what is necessary shall be done". We must avoid the mistake of thinking that Quality Standards are absolute axioms.

2.2 MANAGEMENT REQUIREMENTS OF THE MEASUREMENT INSTRUMENTATION IN QUALITY CERTIFICATION

In the standardization regarding Quality Systems, measurements compatibility is one of the characteristics carefully considered; even more, respect other Corporate Functions that might characterize, more specifically, the product's quality. While the Quality specifications of a product are various and have to be defined case by case, in the measurement field we have the advantage to establish the Quality level in an undifferentiated way. For all types of measurement the relevant uncertainty can be considered a very general Quality index. This characteristic is well defined by BIPM JCGM 100:2008 - Evaluation of measurement data – Guide to the expression of uncertainty in measurement, taken by the ISO Guide 98-3[4] and, in an even more practical way by ISO 14253-2[5].

To summarise the main statements for Quality, starting from ISO 9000, we can see at point 2.3 how the first action required is a clear definition of functional characteristics, that is, the customer needs:

> ***2.3 Quality management systems approach***
>
> *An approach to developing and implementing a quality management system consists of several steps including the following:*
>
> *a) determining the **needs and expectations of customers** and other interested parties;*
>
> *b) ....*
>
> *c) ....*

---

[4] ISO/IEC Guide 98-3:2008 Uncertainty of measurement -- Part 3: Guide to the expression of uncertainty in measurement (GUM:1995)

[5] ISO 14253-2:1999, Geometrical Product Specifications (GPS) -- Inspection by measurement of workpieces and measuring equipment -- Part 2: Guide to the estimation of uncertainty in GPS measurement, in calibration of measuring equipment.

It is evident that customer needs shall be translated, for a production process, in the relevant tolerances according to the scheme of fig. 2.1. Measurement has an important role, being a main part of the third action of a "Process Based Quality Management System": Plan-Do-Check-Act (PDCA). Moreover, measurement and metrology are the object of the part 3.10, some of its clauses being cited here:

> ***3.10 Terms related to quality management for measurement processes***
>
> *3.10.1 measurement management system: Set of interrelated and interacting elements necessary to achieve metrological confirmation (3.1 0.3) and continual control of measurement processes (3.1 0.2).*
>
> *3.10.2 measurement process: Set of operations to determine the value of a quantity.*
>
> *3.10.3 metrological confirmation: Set of operations required to ensure that measuring equipment (3.1 0.4) conforms to the requirements (3.1.2) for its intended use.*
>
> *Note 1 Metrological confirmation generally includes calibration or verification (3.8.4), any necessary adjustment or repair (3.6.9), and subsequent recalibration, comparison with the metrological requirements for the intended use of the equipment, as well as any required sealing and labelling.*
>
> *Note 2 Metrological confirmation is not achieved until and unless the fitness of the measuring equipment for the intended use has been demonstrated and documented.*
>
> *Note 3 The requirements for intended use include such considerations as range, resolution and maximum permissible errors.*
>
> *Note 4 Metrological requirements are usually distinct from, and are not specified in, product requirements.*

Therefore, we can stress again, as ISO 9000[6] says at point 3.10.3 that measuring equipment shall conform to the customer requirements, not obliged to have a formally defined condition. For instance, like it has been well stated in the note 1, metrological confirmation generally includes calibration, but this is not mandatory: an adequate verification (for instance of reproducibility when the task is only comparison) is sufficient. ISO 9001:2008[7] is also careful in referring requirements to real needs, smoothing clauses that could seem imposing absolute rules. For example in ISO 9001:1994 and 9002:1994, the point 4.11 clearly affirmed the necessity of evaluating the uncertainty for every measurement or test result:

> *Test, measurement and experiment equipment has to be controlled in order to insure that their measurement uncertainty is known and compatible with the required measurement needs.*

---

[6] ISO 9000:2005 -- Quality management systems -- Fundamentals and vocabulary

[7] ISO 9001:2008-- Quality management systems -- Requirements

In the new revision, the same requirement of ISO 9000:2008 is more soft:

> *Where necessary to ensure valid results, measuring equipment shall*
>
> *a) be calibrated or verified, or both, at specified intervals, or prior to use, against measurement standards traceable to international or national measurement standards*

but we must remember that the need of evaluating uncertainty remains, as was explained at Chapter 1: the uncertainty evaluation is implicitly necessary.

2.3 MEASURE AND TEST ACTIVITIES MANAGEMENT REQUIREMENTS

The passage from the certification's field to the accreditation one moves the attention from the organizing and system areas to a more complete vision, involving also the final measurement and test results.

It's not enough to limit the control to the quality of the measuring instruments. The possible error causes can't all be attributed to the instrument. Even the best, if used incorrectly, will probably generate wrong results. The correctness of the result can only be guaranteed considering all the components of what is often called "measurement complex". This is the combination of instrument, operator, environment in which the measurement is done and measurand, that is what has to be measured.

It is important to take note of how the measurand can have an important influence on the measurement results; therefore it is incorrect to idealize our measurand. For example, if we have to characterize a ball of a ball-bearing and we consider it as a sphere, we would measure only a single diameter, because for a sphere each diameter is equivalent. But, if we measure the ball with a LVDT displacement transducer having a resolution of 0.1 μm, we can see significant differences in its diameters. Therefore, thinking that the ball of a ball-bearing is a sphere is not correct, and gives us a wrong idea about what we are measuring. A more effective idea of what we are measuring is to think that our ball is like a potato. No one would characterize a potato by a single diameter!

This "Potato theorem" shows us how it is important to consider our measurand in all its characteristics, including possible irregularities.

The correct management of the measure and test activities, fundamental in calibration and testing laboratories, requires, therefore, a global vision, as described in ISO/IEC 17025:2005[8], General requirements for the competence of testing and calibration laboratories. It contains the indication for correctly managing the laboratory activities:

> *1.1 This International Standard specifies the general requirements for the competence to carry out tests and/or calibrations, including sampling. It covers testing and calibration performed using standard methods, non-standard methods, and laboratory-developed methods.*

---

[8] ISO 17025:2005, General requirements for the competence of testing and calibration laboratories.

It is expressly said that the evaluation has to be done on the laboratory competences, involving in such control the available instrumentation and its traceability to the national standards of measurement units, the environmental conditions and the operator's experience. The complete evaluation of the competence comes to examine the global ability with the experimental comparison (audit) of the results that the laboratory is able to obtain to the corresponding results obtained by the Accrediting Authority.

Here we can separately consider the effects of measuring instrument, operator, accommodation and environmental conditions and measurand.

2.3.1 Management of measuring instruments

To highlight typical problems, frequently present in the laboratory activity, it can be useful to describe what really happened in the 50s in the framework of hardness measurement. A scientist studied the causes of errors in Rockwell C (HRC) measurements, showing that the possible difference in results could be up to 1 HRC.

Some producers of hardness testers smiled. A survey of the various good manufacturers of hardness standard blocks (material samples used for calibrating hardness testers) was quoted.

A lot of detailed checks, carried out carefully in their laboratories, and many results of internal tests were reported. No differences higher than ± 0.2 HRC were present. Producers mocked the scientist analysis defining it as "a simply fantastic table of errors"[9].

For a long time the question remained undefined until, in the 90s, in the United States was proposed to use organically hardness measurements for evaluating the tensile strength of screws, bolts, rivets and other similar joint systems.

Safety problems were immediately apparent, because the rupture of a junction (think of the extended riveting of aircrafts) can produce serious trouble. Therefore, the National Institute of Standards and Technology (NIST) picked up the topic and asked to make a series of comparative tests[10].

The results showed that those labs that were convinced of being able to provide results within ± 0.2 HRC were wrong. The real differences between them came up to ± 1.4 HRC.

This case shows that even the work of more thorough check carried out only within the laboratory itself cannot guarantee the goodness of the results. It is, therefore, necessary to trace the laboratory results to an external reference on which they can be compared, as explicitly required, "When necessary", by the ISO 9001:2008 and even in greater detail in the following section of ISO 17025:

---

[9] Small L., Theory and practice of Hardness Measurement, Service Diamond Tool, Ferndale, Michigan, p. 424

[10] Shives T. R., Smith J. H.; Intercomparison study of Rockwell C Scale Hardness Test Blocks; NIST Report NISTIR 4531; 1991

5.6.2.1.1 For calibration laboratories, the programme for calibration of equipment shall be designed and operated so as to ensure that calibrations and measurements made by the laboratory are traceable to the International System of Units (SI).

A calibration laboratory establishes traceability of its own measurement standards and measuring instruments to the SI by means of an unbroken chain of calibrations or comparisons linking them to relevant primary standards of the SI units of measurement. The link to SI units may be achieved by reference to national measurement standards. National measurement standards may be primary standards, which are primary realizations of the SI units or agreed representations of SI units based on fundamental physical constants, or they may be secondary standards which are standards calibrated by another national metrology institute. When using external calibration services, traceability of measurement shall be assured by the use of calibration services from laboratories that can demonstrate competence, measurement capability and traceability. The calibration certificates issued by these laboratories shall contain the measurement results, including the measurement uncertainty and/or a statement of compliance with an identified metrological specifications (see also 5.10.4.2).

The first step to avoid incompatibility of the results is, as required both by the general laws of Quality Systems and by ISO 17025, the use of the traceability to the International System of Units (SI).

Such traceability has to reach the working instruments and it is organized, for practical reasons, in external and internal parts. For the external Company the traceability is managed by the organization responsible for the dissemination of SI, which starts from the Meter Convention[11] (supranational coordination body based on a diplomatic agreement between the Member States).

The diplomatic part of the agreement between the various States is carried out by the General Conference on Weights and Measures (GIPM), while the most scientific part is handled by the International Committee on Weights and Measures (CIPM). The necessity of these two types of bodies includes the interaction between issues specific to the various States and general scientific matters.

For example, the possibility of measuring the mass in a simple way is related to an agreement on the value of air density to be taken as a reference. That density is related to the air pressure reference.

The current definition of the average air pressure at sea level does not give any problem of scientific nature, but entails practical problems for those countries, such as Mexico, where most of the metrological and trade activities take place at high altitudes. It is clear, therefore, that it is appropriate to establish an interaction between a diplomatic context, in which representatives have the rank of ambassador, and a scientific one. The latter, then, is expressed in a specific manner for the various physical quantities through the Consultative Committees. Decisions are made taking into account the demands

[11] Meter Convention. Information on www.bipm.org/en/convention/

brought by the different Member States and the results of scientific experiments carried out in the National Metrology Institutes (NMIs) (especially in collaboration with the international office organized as part of the Meter Convention, namely the Bureau International des Poids et Mesures, BIPM).

2.3.1.1 Dissemination of the International System of Units (SI)

The required traceability to the SI, necessary to guarantee the compatibility of the measurement results, is realized thanks to the activities of dissemination of the National and International Metrological Authorities. To describe such activities we will refer specifically to what happen to the mass.

The International Kilogram Prototype is the starting point. It is a platinum-iridium cylinder (Fig. 2.7), maintained at the BIPM, that has the task to represent the mass unit. The information contained in it has been transferred through a mass comparator with an uncertainty of just few micrograms, to the six official copies (Fig. 2.8).

Other copies have been made, which have been also compared with the prototype through a special mass comparator and distributed to the NMIs. These copies are used in these Institutes to calibrate the stainless steel working copies.

NMIs are responsible for disseminating the units of measurement to users, but it is clear that they are not able to calibrate every weighing instruments used in various industries. So in most of the States, national accreditation body accredit laboratories, with proper technical and organizational ability, such as Calibration Laboratories. These act as intermediaries between the NMIs and Companies in the transfer of the reference to the National Mass Standard, using their masses, calibrated by the NMIs to calibrate, in their turn, the masses and the scales used at industrial level.

Fig. 2.7
The international kilogram prototype conserved under three protective glass bells.

Fig. 2.8
The international kilogram prototype together with its six official copies.

In general, the dissemination of other International Units of Measurement is analogous to the one described before. It is constituted by a metrological chain of subsequent calibrations: the International Standards are used to calibrate the National's ones, which are used to calibrate those owned by the Calibration Laboratories which, finally, provide the first connection used within the Company.

In the case the international standards do not exist, the national standards, normally realized by the NMIs through primary methods, are compared and mutually recognized by the states signatories the Mutual Recognition Arrangement (MRA) that give the rules since this mutual recognition can be assured[12]. The schematization of dissemination, as described in Fig. 2.9, includes the following:

1. the international level, realized by the Meter Convention for scientific and diplomatic part through the Mutual Recognition Arrangement (MRA) and, in Europe, by EA[13] (European cooperation for Accreditation) for the quality management and the mutual recognition of the national accreditation activities in Europe;
2. the national level , in which National Metrology Institute (MNI) operate to maintain and compare the International System of Units (SI) and to make it available for the dissemination;
3. the level of calibration laboratories, established in almost every country of the European Union through Accreditation Bodies (e.g. ACCREDIA[14] in Italy, Deutsche Akkreditierungstelle GmbH (DAkkS)[15] in Germany, UKAS[16] in United kingdom, COFRAC[17] in France), which act as intermediaries between the NMIs and users for the transfer of metrological references. Users can therefore have some instruments connected directly to the international metrological organization, called first line standards;
4. the dissemination continues within the Company with a similar pyramidal scheme, which starts from the first line Company Standards used to calibrate the working instruments. Eventually, if these are too numerous, through second line Company standards.

The dissemination of the SI, corresponding to the traceability request of measuring instruments to the National Mass Prototypes, is a necessary condition to ensure the compatibility of measurement results and test among different laboratories.

As at the international level the MRA assure the mutual agreement between the calibration certificates issued by the NMIs (e.g. in Europe, INRiM[18]

---

[12] BIPM, Bureau International des Poids et Mesures, Pavillon de Breteuil, Sèvres. Information on www.bipm.org/en/cipm-mra/

[13] EA, European co-operation for Accreditation. Information on www.european-accreditation.org/

[14] ACCREDIA, L'Ente Italiano di Accreditamento. Information on www.accredia.it

[15] Deutsche Akkreditierungstelle GmbH (DAkkS) http://www.dakks.de/

[16] UKAS, United Kingdom Accreditation Service. Information on www.ukas.org

[17] Cofrac - Comité français d'accréditation www.cofrac.fr/

[18] INRIM, Istituto Nazionale di Ricerca Metrologica. Information on www.inrim.it

and METR[19] in Italy, PTB[20] in Germany, NPL[21] in United Kingdom, LNE[22] in France), in Europe the mutual agreement under EA is such that any NMI calibration certificate issued by any European Accredited Laboratory can be mutually recognized as shown in fig. 2.9.

2.3.1.2 SI dissemination within Companies

Within Companies it must be ensured, as required by ISO 9001:2000, the traceability of working instruments to the SI. It is clear that the Calibration Laboratories cannot calibrate all the working instruments used by end users.

Even within the Company there must be a dissemination of their measurement units with a pyramid scheme (Fig. 2.9), when the number of working instruments is so large that the Company First Line Standard is not sufficient for doing all the required calibrations.

To continue with the example of the mass, the Company, if needs it, will buy a set of Standard Masses and will ask a Calibration Laboratory for calibration, at a Quality level appropriate to their needs. That set of masses will be the Company First Line Standard, its starting point for the internal dissemination (indicated by the letters CApl in Fig. 2.9).

The Company will then evaluate the dissemination pattern more efficient from a practical and economic standpoint. We get the examples in fig. 2.9 in the Company 1 description:

- a quite common situation is to use a First Line Standard to calibrate directly the working instruments (see pattern CApl2,, SL2);
- when the number of working instruments is too large for a single Standard, it is possible to calibrate with the First Line Standard one or more Second Line Standards (see pattern CApl1, CAsl1, Sl1-Sl2);
- when the number of working instrument to be calibrated with a given type of standard is small, it is not convenient to set a calibration line installing a First Line Standard, but is more efficient to ask directly the calibration to a Calibration Laboratory (pattern ACCREDIA3, SL3).

It is very important to evaluate the alternatives based on their efficiency, proposing clear goals (ISO 9001:2000 and ISO 17025), and estimating the costs corresponding to the different alternatives. Note that, while outside the Company traceability of the calibration is guaranteed by authorities that manage the dissemination of the SI, within the Company traceability must be guaranteed with a careful evaluation of uncertainty for each step of the metrological chain. In fact, the compatibility of results is obtained if the uncertainty is calculated correctly, in order that the uncertainty interval contains the reference value (Fig. 2.10).

---

19 Istituto nazionale di metrologia delle radiazioni ionizzanti - METR - ENEA, http://www.enea.it/it/chi-siamo/documenti/struttura-organizzativa/metrologia.pdf

20 PTB, Physikalisch technische Bundesanstalt. Information on www.ptb.de

21 NPL, National Physical Laboratory. Information on www.npl.co.uk

22 LNE, Laboratoire national de métrologie et d'essais. Information on www.lne.fr

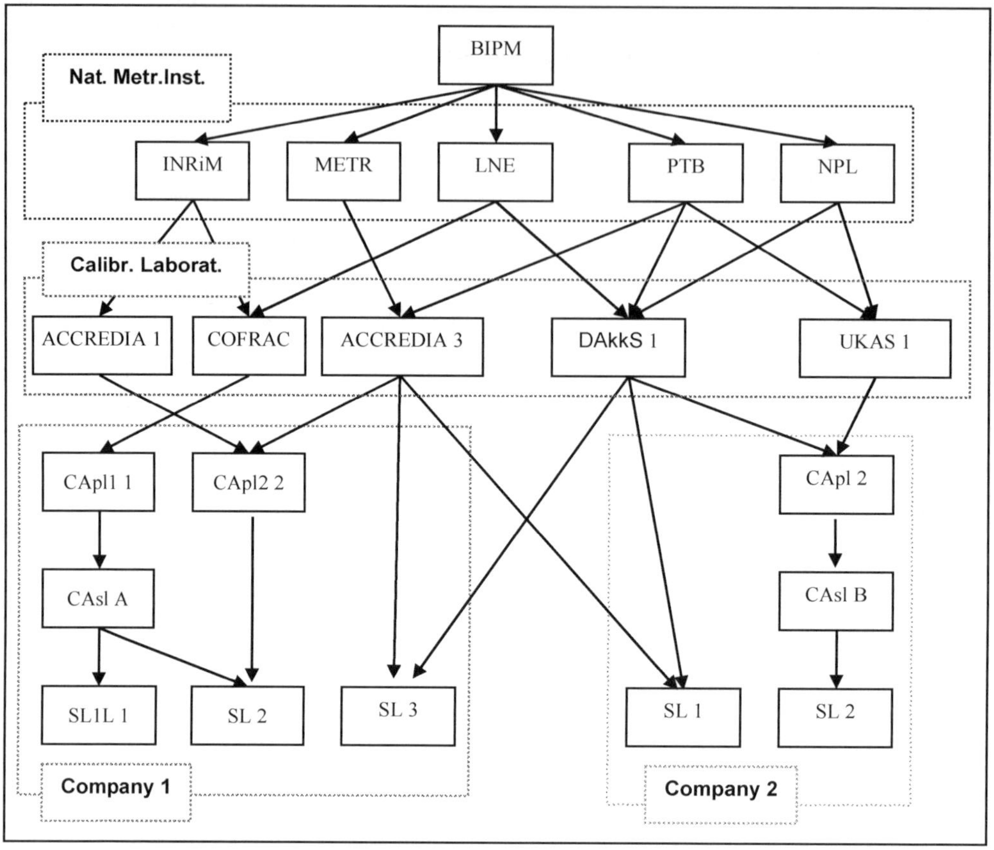

Fig. 2.9 *Dissemination of SI in Europe, starting from the Bureau International des Poids et Mesures (BIPM), trough MNIs and accredited Calibration Laboratories to provide Calibration Certificates to Industry traceable to the SI.*

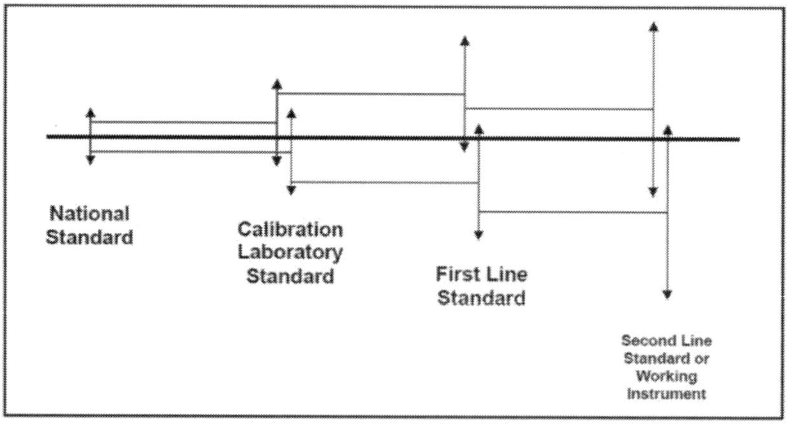

Fig. 2.10 *Importance of uncertainty to manage the traceability chain and the compatibility of the results.*

The working instruments, even if calibrated independently with different metrological chains, will have uncertainty intervals that will overlap, at least at the point corresponding to the reference value. So the condition of compatibility will be automatically assured.

In practice, the passages from the First Line Standard to the working instruments must be carefully documented with written technical procedures, which define the calibration to be carried out and express the corresponding uncertainty.

Traceability of measuring instruments produces two basic characteristics:

- a reasonable confidence that the scale of the instrument is free from errors of sufficient magnitude to make the results inadequate to the purposes;
- the evaluation of the uncertainty interval only due to the instrument.

Obviously, this does not complete the necessary information in order to have all due guarantees on the provided results. In fact, the result does not depend only by the instrument, but also on the circumstances in which it is used, as shown in the technical part of ISO 17025:

> ***5.1.1*** *Many factors determine the correctness and reliability of the tests and/or calibrations performed by a laboratory. These factors Include contributions from:*
>
> *- human factors (5.2);*
>
> *accommodation and environmental conditions (5.3);*
>
> *test and calibration methods and method validation (5.4);*
>
> *equipment (5.5);*
>
> *measurement traceability (5.6);*
>
> *sampling (5. 7);*
>
> *the handling of test and calibration items (5.8).*
>
> ***5.1.2*** *The extent to which the factors contribute to the total uncertainty of measurement differs considerably between (types of) tests and between (types of) calibrations. The laboratory shall take account of these factors in developing test and calibration methods and procedures, in the training and qualification of personnel, and in the selection and calibration of the equipment it uses.*

It is, therefore, necessary to organize a complete scheme of the measurement and test management, taking into account, at least, the staff experience and the environmental conditions.

2.3.2 Staff experience and training

The operator influence on the measurement results is usually associated to a condition linked to the random reading errors. In fact, scientific studies have shown that the influence can also be systematic in the sense that an operator can always give larger measurements (or smaller) than the average of the others. The Japanese[23] even came to do the calibration of the operators, giving each of them a correction factor.

Regarding the staff, the ISO 17025 says:

> ***5.2.1*** *The laboratory management shall ensure the competence of all who operate specific equipment, perform tests and/or calibrations, evaluate results, and sign test reports and calibration certificates. When using staff who are undergoing training, appropriate supervision shall be provided. Personnel performing specific tasks shall be qualified on the basis of appropriate education, training, experience and/or demonstrated skills, as required.*

As we can see the requirement has a very practical scope. In the field of measurements the operator's experience is often crucial, especially for those parts that can hardly be standardized or fully described in a technical procedure. In quite correlated field of work there are specific standard documents on personnel that can be useful. We can indicate one used in a typical field in which the staff experience is well defined: the non-destructive testing (EN 473 / ISO 9712 Non-destructive testing -- Qualification and certification of NDT personnel):

> ***NOTE 1*** *In some technical areas (e.g. non-destructive testing) it may be required that the personnel performing certain tasks hold personnel certification. The laboratory is responsible for fulfilling specified personnel certification requirements. The requirements for personnel certification might be regulatory, included in the standards for the specific technical field, or required by the customer.*

As usual, we must not consider the situation as ideal, but it has enough common sense to understand what might be the effect of the operator and, if necessary, to make evaluation tests (i.e. try to estimate the measurement difference between different operators).

2.3.3 Accommodation and environmental conditions

The accommodation and environmental conditions, mainly the temperature, but sometimes also the atmospheric pressure and the relative humidity, may affect the measurement results. Also other factors, like the oscillatory phenomena, both mechanical (vibration) and electrical (electro-magnetic interference), have behaviours that can be unexpected. We have seen, for example, the behaviour of a very sensitive (10^{-6}) electronic instrument. Despite the high sensitivity it was a very "robust" indicator: the reading moving no more than one digit during measurement in a workshop with machine tools and arc welding devices in action but meanwhile I was measuring, the ringing of an old telephone made the instrument go crazy, producing for a while a strong and irregular variation of hundreds digits. This example shows that it is

[23] Hida N., Yamamoto K., "On the determination of absolute value of diagonal length of Vickers indentation". Bulletin of the National Research Laboratory of Metrology No 25, 1970

advisable to check that the electrical equipment do not release too intense electromagnetic disturbances and measuring instruments are immune to receive those or claiming to have received them, refusing to provide an indication of the measurement in such conditions.

About the environment, the ISO 17025 says:

> ***5.3.1*** *Laboratory facilities for testing and/or calibration, including but not limited to energy sources, lighting and environmental conditions, shall be such as to facilitate correct performance of the tests and/or calibrations.*
>
> *The laboratory shall ensure that the environmental conditions do not invalidate the results or adversely affect the required quality of any measurement. Particular care shall be taken when sampling and tests and/or calibrations are undertaken at sites other than a permanent laboratory facility. The technical requirements for accommodation and environmental conditions that can affect the results of tests and calibrations shall be documented.*

Also for managing the line of requiring "what is need" of the recent ISO Standards is certainly better than the rigid requirements of the old ones. Before, one of the points of debate concerned the management of measuring instruments used to evaluate environmental conditions. Even when faced with large tolerance of temperature, e.g. from 18 °C to 28 °C, range very easily controlled with a simple and non-expensive thermometer, with the old standards was sought calibration and certification assuring traceability to the SI, which often cost more than the thermometer itself. The new ISO Standards allow, however, to evidence the real needs and use, where appropriate, simplified procedures as described in Appendix 2.1.

2.4 MEASUREMENT PROCEDURES

A strong influence on the results is due to the how the measurement process is organized. In Quality, it must be described by technical procedures able to apply it in all its parts including, if necessary, the sampling of specimens parts that need to be measured and their manipulation (Section 5.1.1. ISO 17025) and the assessment of uncertainty, which will determine the compatibility condition:

> ***5.4.1 General***
>
> *The laboratory shall use appropriate methods and procedures for all tests and/or calibrations within its scope. These include sampling, handling, transport, storage and preparation of items to be tested and/or calibrated, and, where appropriate, an estimation of the measurement uncertainty as well as statistical techniques for analysis of test and/or calibration data...*

We note immediately that in this standard has been paid much attention to en-sure that it will not be possible to read the directions given as absolute impositions. It is, in fact, expressed a clear correlation between things to do and corresponding tar-gets. Very important is the attention to not take unnecessary paperwork, expressed in a note in the point 5.4.1:

NOTE International, regional or national standards or other recognized specifications that contain sufficient and concise information on how to perform the tests and/or calibrations do not need to be supplemented or rewritten as internal procedures if these standards are written in a way that they can be used as published by the operating staff in a laboratory. It may be necessary to provide additional documentation for optional steps of the method or additional details

I wanted to mention this principle first, because it is often overlooked either in the direction of the inadequacy, either in the direction of unnecessary redundancy. The case of extreme care in the preparation of the paper and the lack of measuring instruments which often becomes absolute in the evaluation of uncertainties is a very common case. It is also not unusual in case of measuring instruments with characteristics too high for the planned measurement's objectives or included in an excessive periodic pattern of calibration. Such conditions can lead on one side to the risk of not allowing the achievement of the objectives, on the other to an excessive cost.

2.4.1 Nonstandard methods

The focus on the well-harmonized management of the procedures according to the need is particularly important in the field of non-standardized tests. These tests are particularly important in the Research and Development areas and to support the design, but they are not rare even in normal testing. Even in this case, it happens very often that the judgment on the compliance or non-compliance of the tested piece should be drawn in a targeted manner. Sometimes combining a series of standardized test procedures, but also adopting an entirely new and specific testing procedure. The management of non-standardized tests, while having to fulfil the same basic requirements of standardized tests, in other words, to ensure the compatibility of the results, it is more complex because in it the number of degrees of freedom is much higher.

In case of a normalized test, for example the measurement of hardness, it is sufficient to follow the indications of the standard: verification of the hardness tester with hardness reference blocks, verifying that the environmental conditions comply with the expected temperature range and instructing the operator to maintain the required loading times and to make a correct reading of the instrument. Even the part related to sampling (selection of the measurement points on the sample) is free and, usually, is not left to the operator, but agreed with the client.

In the case of non-standardized tests, however, the test plan must be completely agreed. This involves so many levels that each of them requires specific expertise. ISO 17025 provides a framework of activities to schedule:

5.4.4 Non-standard methods

When it is necessary to use methods not covered by standard methods, these shall be subject to agreement with the customer and shall include a clear specification of the customer's requirements and the purpose of the test and/or calibration. The method developed shall have been validated appropriately before use. NOTE For new test and/or calibration methods, procedures should be developed prior to the tests and/or calibrations being performed and should contain at least the following information:

a) appropriate identification;

b) scope;

c) description of the type of item to be tested or calibrated;

d) parameters or quantities and ranges to be determined;

e) apparatus and equipment, including technical performance requirements;

f) reference standards and reference materials required;

g) environmental conditions required and any stabilization period needed;

h) description of the procedure, including

- *affixing of identification marks. handling, transporting, storing and preparation of items,*
- *checks to be made before the work is started,*
- *checks that the equipment is working property and, where required, calibration and adjustment of the equipment before each use,*
- *the method of recording the observations and results,*
- *any safety measures to be observed;*

i) criteria and/or requirements for approval/rejection;

j) data to be recorded and method of analysis and presentation;

k) the uncertainty or the procedure for estimating uncertainty.

To determine what suggested by the standard it is necessary to: make a hypothesis on the test parameters that realize the goal, plan ways to determine and validate the test methodology so defined. It is clear that the choice of parameters to be measured requires to have jointly the expertise to define the subject of the test, its relations with measurable parameters and metrological competence to properly plan the measurements. It is evident how the validation methodology for a normalized test often consists in the repetition of the test under different conditions. These hold a reasonably representative field of the possible situations. This is only possible because the amount of work involved is justified by the hundreds or thousands of tests performed then. In case of the not normalized tests, this would be economically impossible; therefore, the validation can often be made only on theoretical backgrounds. A perfectly parallel considerations made on the validation must be done for the uncertainty assessment.

2.5 CONCLUDING REMARKS OF GENERAL NATURE

The UNI EN 45001, now replaced by ISO 17025, said, in the old revision, that in the calibration certificates and test reports no judgment could be expressed. It was carried to the point that, in practice, the Accreditation Bodies in in Italy did not allow to declare, along with the results, the compliance with the Standard Specification that had been used to do the tests. For example, the calibration certificate of a materials testing machine had to contain all the results imposed by ISO 7500-2, but was forbidden the next step, the classification of the machine following the reference standard. This consists in comparing the calculated values with the standard requirements verifying in which class the tested machine fell.

The new ISO 17025 will examine the problem with more practical sense:

5.10.5 Opinions and interpretations

When opinions and interpretations are included, the laboratory shall document the basis upon which the opinions and interpretations have been made. Opinions and interpretations shall be clearly marked as such in a test report.

NOTE 1 Opinions and interpretations should not be confused with inspections and product certifications as intended in ISO/IEC 17020 and ISOIIEC Guide 65.

NOTE 2 Opinions and interpretations included in a test report may comprise, but not be limited to, the following:

- *an opinion on the statement of compliance/noncompliance of the results with requirements;*
- *fulfillment of contractual requirements;*
- *recommendations on how to use the results;*
- *guidance to be used for improvements.*

NOTE 3 In many cases it might be appropriate to communicate the opinions and interpretations by direct dialogue with the customer. Such dialogue should be written down.

I mentioned this topic on opinions and interpretations because, in the past, has raised many discussions and emphasizes once again how the evolution of the regulatory environment has shifted the focus of attention from a formalism, often only on paper, to substantial requirements.

To conclude this brief presentation, I want to highlight the fact that only the main problems concerning the management of a laboratory were treated. ISO 17025, in reality, is very rich in information concerning all steps an organization, also at very practical level.

APPENDIX 2.1. THE ECONOMIC MANAGEMENT OF TRACEABILITY

The problem to face situations inadequate to the management of the commonly used tools is not uncommon. Apart from the excessive cost, we run the risk of being placed in front of a refusal of a Calibration Laboratory for formal reasons (their test procedure expects to be applied only to instruments of a certain class), or for substantial reasons (the application of a calibration process to an inadequate instrument may damage the reference standard).

An example may concern the accessory instruments, those that are not used to measure directly the interest quantity, but rather the influence quantities. A typical condition in mechanical measurements is the request that the operating temperature has to be within the range from 18 °C to 28 °C. This can be satisfied with low cost thermometers.

The first, valid especially when the required instruments have average precision, is to purchase an instrument with the manufacturer's declaration of conformity in accordance to our needs and in establishing regular extra-long calibration intervals. These can be accepted by the Certification authority for the high level of uncertainty, appropriate in these cases. In practice we can reach intervals of recalibration corresponding to the expected life of the instrument.

If the tolerances to be observed are more restricted, such as those required for length measurements already refined (i.e. from 18 °C to 22 °C) or if we think we cannot convince the Certification Authority to accept very long recalibration intervals, it may subject the operation of recalibration to the outcome of intermediate verification procedures provided by the ISO 10012:2003 - Measurement management systems -- Requirements for measurement processes and measuring equipment. This intermediate verification consists of operations that, although not able to ensure the traceability to national standards, allow to highlight significant irregularities of operation. In case of accessory instruments it may be sufficient a comparative reproducibility test.

The third solution is the classic one, which consists in having a thermometer calibrated periodically to ensure an uncertainty of less than a few tenths of celsius degree and compare it with the other environment control thermometers with a procedure similar to the one described before,. The proposed procedure allows to qualify these thermometers in their range of normal use, so it is adequate to ensure the traceability with the uncertainty of some tenths of a celsius degree that may be required.

Do not practice the solution consisting in the acquisition of an apparatus for calibrating thermometers (with a thermostatic chamber and a standard thermometer) which would create an unnecessary decrease of uncertainty and, in this case, would highly increase the costs.

3 CHARACTERISTICS OF MEASURING INSTRUMENTS

3.1 INTRODUCTION

General rules, like the ones examined in the previous chapter, allow a good schematization of the system, but cannot be too specific. It is, therefore, important to remember that the scheme has to correspond to reality and to be completed in the details. We go back to what said in the ISO Guide to the expression of uncertainty in measurement, already quoted in the preface:

> *"Although this guide provides a framework for assessing uncertainty, it cannot substitute for critical thinking, intellectual honesty and professional skill. The evaluation of uncertainty is neither a routine task nor a purely mathematical one; it depends on detailed knowledge of the nature of the measurand and of the measurement".*

The points to consider, other than the most important one, the intellectual honesty (that has a precise moral meaning), are the critical thinking and professional skill. With the more technical part of the uncertainty evaluation, that collects the deepest and more intimate essence of measurement, these are all together in the *"detailed knowledge of the nature of the measurand and of the measurement"*.

In fact, many, if not even all choices that has to be done cannot ignore specific information. Let's quote some example.

In Rockwell C hardness measurement a diamond indenter, shaped as a cone with 120° of opening angle, is used. The rounded point is a spherical cap having 200 µm of radius. Tolerances are quite strict and its measure is complex, being a really small object. A hypothesis of procedure could be based on the determination of the coordinates of a certain amount of points on the indenter surface and the relative use of the statistical method to identify from these points the general geometry. Such operation can be organized in two different ways:

- separated determination of the spherical and the conical parts of the geometry;
- use, as geometrical model, a segmented function that can describe the two parts together.

These two methods of analysis give completely different results. While the second one unifies well the two geometrical components (it contains a defined relationship between position and tangency on the common border), the first leaves them free to have any relative position, and therefore, on the border there is a discontinuity that does not exist in reality. It might seem immediate the choice of the second method as more adapt, but it is not as easy as it seems.

It is important to understand if the conical and spherical parts were born constructively connected. In this case the second method is more correct. If, instead, they are connected as a finish, being born separately, then the first method better represent reality.

Only the knowledge of the construction process can solve this dilemma. If the processing phases of the cone and the sphere are independent (i.e. made with different machines), then the first method is more correct and the shift value of the spherical centre from the axis of the cone can be significant to understand the indenter behaviour. If the production of the two parts is unitary, then the second method is the most correct one and it avoids that random errors[1] lead to highlight not significant shifts between the two geometrical parts.

As second example we can examine a circularity test done measuring a set of diameters of a nominally circular section with a digital micrometer and evaluating the differences between all these measurements.

The circularity[2] tolerance can be defined like the difference between the diameter of a circumscribed circumference and the diameter of an inscribed one in the section under consideration, in order to delimit the shape irregularities. Talking about circumferences (circumscribed and inscribed) it indicates immediately that, for the circularity control, the correspondent centres have to be individuated. The measurement done with a digital micrometer takes into account only the extreme points of a series of diameters, leaving completely free the positions of their central points.

Let's examine, in a completely theoretical way, what could happen using two gage plug having the same circularity error: the first having oval section (Fig. 3.1) and the second having trilobate section (Fig. 3.2); all cases that can be verified practically.

The values of radii and diameters, as theoretical estimate of the measures taken with a digital micrometer, are reported in table 3.1.

We can observe that, if we take into account the radius values of both shapes, oval and trilobate, we can see that the difference between the diameter of the circumscribed circumference (20.020 mm) and of the inscribed one (19.980 mm) is 0.040 mm.

In case, instead, we decided to use the values of the diameters, like they should be measured using a digital micrometer, we can find that for the oval shape the difference between maximum and minimum diameter is still 0.040 mm, while for the trilobate one such difference is null.

We can see that the measurement taken with a digital micrometer does not allow, with such measuring, to highlight any circularity error.

It is clear, therefore, how important is to know the measuring characteristics to be able to choose the correct measurement method.

---

[1] The term error, widely used in the past, is now questioned in the field of metrology. It is used here, however, by defining it as "difference between the measurement result and the reference value". In practice, as the difference between the value obtained and the one that can be obtained applying the best equipment and measurement techniques available (primary standards).

[2] ASME Y14.5 – 2009. Dimensioning and Tolerancing, ISO 14405-1:2010 -- Geometrical product specifications (GPS) -- Dimensional tolerancing -- Part 1: Linear sizes and ISO 14405-2:2011 -- Geometrical product specifications (GPS) -- Dimensional tolerancing -- Part 2: Dimensions other than linear sizes

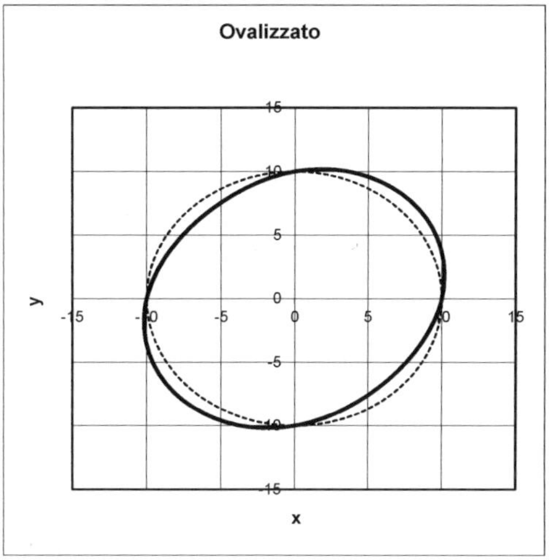

Fig. 3.1
Scheme of a circular oval section

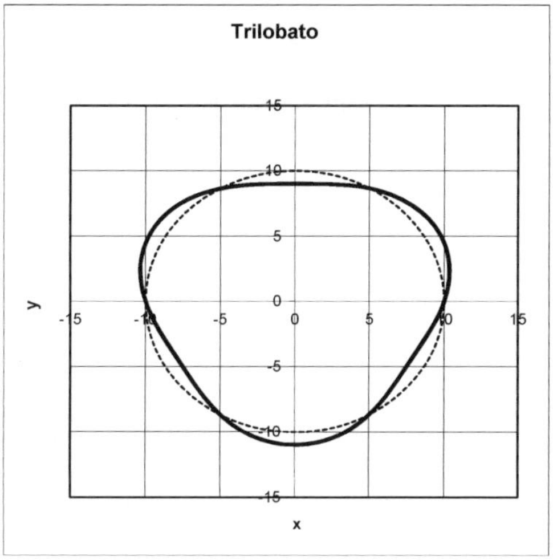

Fig. 3.2
Scheme of a circular trilobate section

Tab. 3.1 Theoretical values of radius and diameter of the oval (elliptic shape) and trilobate gage plugs. Note how the measure of the diameters cannot detect the trilobate shape.

Angle	Oval		Trilobate	
	Radius/mm	Diameter/mm	Radius/mm	Diameter/mm
0°	10,000	20,000	10,000	20,000
15°	10,005	20,010	10,007	20,000
30°	10,009	20,017	10,010	20,000
45°	10,010	20,020	10,007	20,000
60°	10,009	20,017	10,000	20,000
75°	10,005	20,010	9,993	20,000
90°	10,000	20,000	9,990	20,000
105°	9,995	19,990	9,993	20,000
120°	9,991	19,983	10,000	20,000
135°	9,990	19,980	10,007	20,000
150°	9,991	19,983	10,010	20,000
165°	9,995	19,990	10,007	20,000
180°	10,000	20,000	10,000	20,000

As third example, we can question ourselves if the test of a ball bearings production could be done with a simple and fast procedure, like the weighting on a balance with adequate metrological characteristics. With a simplified model (deterministic model of the exact shape) it is possible to calculate that a sphere of 5 mm of diameter has a mass of about 4.14 g.Using a balance with quite normal characteristics, having resolution of 0.1 mg, we should have a relative resolution on the mass of about $2.5 \cdot 10^{-5}$. Therefore, on the diameter, we will have a relative resolution of about $8 \cdot 10^{-6}$, corresponding to about 0.04 μm, well adequate for the dimensional test required. The practical possibility to apply this method can be decided only depending on the detailed knowledge of the production process. This method, in fact, only produces an estimate of the average diameter value. It is not able to individuate the presence of shape errors and it is sensitive to the density variations of the material. Given examples indicate how necessary is to realize that the achievement of the real objectives always requires a deep knowledge of the problem.

This implies even how necessary is a tight collaboration with people having a specific experience. Often the same operators, even without a deep cultural base, developed with the experience a precious sensibility to the measurement conditions. It is important, therefore, to be ready to accept all contributes and use them for their value. It is also important to remember that the task of who organizes the measurement is fundamental. The only experience limited to the practical work on measurement, even if decennial, if not organized in a rigorous way, can lead to misleading results.

In this chapter we will see the principal metrological characteristics of a measurement complex, in order to give us the necessary elements to understand even better all the schemes adopted by the standards ISO 9001 and ISO 17025, already described.

3.2 THE METROLOGICAL CHARACTERISTICS OF MEASURING INSTRUMENTS

In dealing the metrological characteristics of measuring instruments, we will make the opportune connections to the problems related to the presence of random factors for determining of the most significant values of a group of measurements and estimating the relevant dispersion ranges.

We will analyse only the metrological characteristics of: resolution, sensitivity, stability, precision (repeatability and reproducibility), trueness, accuracy and uncertainty. These are useful to allow the measuring activity to be done correctly and, at least, some of them are strictly connected to statistical concepts.

Before focusing the attention to the metrological characteristics themselves, it is important to repeat how, although theoretically they could be attributed to the measuring instruments (therefore evaluated in the calibration), in practice they result strongly connected to the entire measurement complex (measuring instrument, operator, environmental conditions and measurand).

Taking into account a measurement activity, we can still remember the previously described correlation uncertainty, which take into account the effects of measurement strategy and method uncertainty, which depends on how the measurement strategy is applied correctly. We always have to evaluate the situation extensively, referring to what can be defined the measurement complex and how it really point to measure the requested quantity (sometimes what is done is a very careful measurement of a quantity that is simply different from what required. Remember the example of fig. 3.2 on determining the circularity error using diameter measurements).

Regarding the environmental conditions, they are considered influence quantities as described in Chapter 2. Even the effects of the vibration and oscillation conditions, important for the mechanical part, and the electromagnetic fields, important for the electronic part, can be highlighted comparing the test results made in normal and more protected measurement conditions (e.g. during the night when the anthropic noise is reduced). Differences in the characteristics of the results obtained in the two different situations can give indications of the effect of disturbs in the real working condition against a controlled one, possible only in limited tests (see hereafter the concepts of repeatability-reproducibility, a base of very common industrial tests).

The effects of the operator and of the measurement procedures have to be evaluated case by case, eventually with experimental evidences that can highlight capacities, defects, advantages and disadvantages.

In the global exam of the situation, it can be quite useful to use a scheme that contains a clear connection to the various metrological characteristics. This allows a *targeted* management, only limited by the determination of the characteristics really necessary for every instrument and measurement complex for the specific task. This is meant to be a way to prevent wastes of resources (human, economic and time).

To perform a complete measurement activity, we have to know all the characteristics till the uncertainty, but frequently a complete measurement is not necessary: as underlined in Chapter 2, ISO Standards on quality are now very keen in asking to take actions "... when necessary ...". The use of a measuring instrument is frequently limited to monitoring stability or for comparison tasks. For example, we have to remind that many of the application of the measurement done at the industrial level are comparison[3] activities. These do not require knowledge of all the characteristics, but only the ones that manage the ability to establish a comparison.

A good way for understanding metrological characteristic is following the line of the actions one can do with a measuring instrument. The first action is to take the measurement information out of the instrument, directly or by computer. In any case we need a writing-reading capability and its characteristic is the ability of giving a detailed information, called *reading resolution*.

After having the possibility to read our instrument, we need that the value we get do not change too rapidly, otherwise it is useless. We need that it remains nearly constant for the time required. This characteristic is called *stability*.

---

[3] Very common in production, to the point that many dimensional testing tools, reset on a sample piece, show on their scale the extreme limits of acceptance. Therefore the operator does not evaluate the measured value, but only if the difference to the sample piece is contained in those limits.

But stability is applied for the time our measurand remain constant, and this is not the general use: frequently the instrument start from zero, go to the measurement level and return to zero. We will know if this cycle produce some effect on the result, therefore we repeat the cycle to see if the result changes. The characteristic of getting nearly constant results in repeated measurements is called *precision* and depending on the way repetitions are done we have the concepts of *repeatability* or *reproducibility*.

Precision allow to know how large are the differences in repeated measurements, but frequently we want to assign a value to our measurand. If our instrument has a systematic error, e.g. our ruler has the indication 121 at the length 120, when we measure a segment long 120 we get the measurement 121. The amount of systematic error against a reference standard (frequently called bias) is considered by the characteristic called *trueness*.

But doing measurements we just will have a result near to the value of the measurand and frequently do not care if the difference is due to random or systematic errors: we are interested in their sum. The capability of having together good precision and small bias is called accuracy. Notice that the characteristics given up to now are indicated by numbers that go in the opposite direction: high precision is referred by the number corresponding to the small random error, the same for high resolution, high stability, high trueness and high accuracy. But direct indications as small bias->small number, large bias->large number is considered better, therefore the term uncertainty, which gives the same metrological characteristic as accuracy, but with direct indications, should be preferred.

3.2.1 Reading resolution

The first indispensable thing that is required to a measuring instrument is to transfer to the user the information acquired with the measurement. In many cases this happens with the reading operation. Let's examine the reading itself. The basics concepts are present in everyone's experience[4] and it can be then moved to the case of automatic data transmission to acquisition systems managed by a machine.

In many cases this transfer of information is due to the reading action of an operator that records what shown by the position of an index on a graduated scale. In that case it is said that the reading is on an analogue scale. The instrument transforms the measured quantity in another one visible to the operator (i.e. dials and hands of a watch). Alternatively, the acquired information is shown by a numeric indicator, also called digital, which can be transferred through reading. However, it is often transferred though a direct connection to other instruments or to a computer.

In both cases we have to pay attention to the value of the smallest division of the indication, also sometimes called *elementary indication* or *unit of format* or simply *division*. For example, the long hand of a watch has as unit one

[4] Every day we have to read many instruments, watch, speedometer, etc.

minute, but when there is the seconds' hand or, in case of a digital watch showing the seconds, the unit of format is one second.

Regarding the choice of unit of format by the builder, many factors need to be taken into account, manly the real capability of resolution of the instrument. We shall remember that VIM gives a definition of resolution:

> *4.1.14 resolution*
>
> *smallest change in a quantity being measured that causes a perceptible change in the corresponding indication*
>
> *NOTE Resolution can depend on, for example, noise (internal or external) or friction. It may also depend on the value of a quantity being measured*

and reading resolution of the indicator scale:

> *4.1.15 resolution of a displaying device*
>
> *smallest difference between displayed indications that can be meaningfully distinguished*

We can see that resolution depends on the real working precision of the instrument (e.g. if there is noise on electronic circuitry of the instrument the output is oscillating and the resolution corresponds to the amount of oscillation) while the reading resolution just depends on reading scale (for instance, in a digital indicator, the elementary interval between indications). We shall consider also that resolution is an important factor for establishing the price of an instrument (an instrument having 1 µm resolution is more expensive then the same type of instrument having 10 µm resolution) and that it is not difficult, for the instrument producer, to reduce the unit of format; it is really possible that commercial data sheets indicates reading resolutions smaller than the real resolution capability of the instrument. To better define this concept, the real factor affecting the price is the ratio of resolution to the *range of nominal indication interval* (VIM 4.5) sometimes called *span*. It is the range within which the instrument gives an output (e.g., a voltmeter scale going from -10 V to +10 V has a range of nominal indication interval of 20 V)

To clarify the way of using that ratio we shall give some examples. Let's start with the example of a delicatessen seller's scale. We are in the legal metrology field, so the resolution corresponds almost completely to the uncertainty value. A typical value for the resolution of this kind of scale is 2 g, because, with such resolution, the law allows to weight the quantities usually required by the customers (i.e. 100 g). However, is important to understand that the seller does not use the balance only when he sells his products, but even when he buys them. In the second case, the quantities are about 10 kg units or more. The most usable balance is, therefore, the one that gives at the same time a resolution of 2 g and a range exceeding 10 kg. Therefore, an important usability characteristic is the ratio between the range to the resolution. This ratio is so important that in some cases the instrument (mainly Legal Metrology instruments, which uses are regulated by law following the

Standard Documents[5] of the International Organization of Legal Metrology, OIML) is defined by that value (e.g. the OIML C3000 category for legal balances defines instruments having the span equal to 3000 units of format). So a normal balance with 3000 divisions could be quite good for a customer (resolution 1 g and 3 kg of range), but it is not usable to weight entire hams. The most common choice is a more expensive balance with 6000 divisions. It penalizes a little the customer, moving the resolution to 2 g, but satisfied the needs of the delicatessen seller, bringing the range to 12 kg.

We have to notice that the classification in divisions can be applied widely: it can regard both small counter balance (4000 divisions, 1 g resolution, 4 kg range) and a weighbridge (4000 divisions, 20 kg resolution, 80000 kg range).

Finally, it is useful, talking about resolution, to underline immediately the importance of not misleading such concept with *sensitivity*, which is the ratio between the value of the output quantity to the input quantity. In balances, like in every direct reading instrument (where input and output quantities have the same measurement units), the sensitivity is always 1.

The concept of sensitivity becomes important both for instruments having different input-output measurement units and when the measurement scales are intrinsically not linear, or even when we want to describe the instrumental sensitivity to the influence quantities (i.e. the sensitivity to temperature, etc.).

In case of a load cell, for example, we have a force as input (let say measured in newton[6]) and an electric tension as output (measured in microvolt). Here we see clearly how a load cell having sensitivity of 3 µV/N is more sensitive than one with just 2 µV/N, or how a balance with a temperature sensitivity of 2mg/°C would suffer more any temperature variation than another one with just 1mg/°C.

In case of instruments with intrinsically not linear characteristics, the sensitivity varies from point to point in the measurement scale. For example, anemometers have usually, as output, a difference of pressure p proportional to the square of the velocity v measured:

$$p \propto v^2$$

being the characteristic known, it is possible to calculate the sensitivity as partial derivative of the output to the input:

$$S_v = \left(\frac{\partial p}{\partial v}\right)_v \propto v$$

therefore, in the examined case, the velocity sensitivity will be proportioned to the velocity itself.

---

[5] WELMEC 2.1 (Issue 4) August 2001-. Guide for Testing Indicators (Non-automatic Weighing Instruments). http://www.welmec.org/fileadmin/user_files/publications/2-1.pdf

[6] All measurement units have inherited their name from persons: from I. Newton, B. Pascal etc. Their symbol has a capital letter (N, Pa, etc.), but when they are written for extended they have to be written with a lowercase letter (newton, pascal, etc.).

Coming back to resolution, for technological instruments in normal conditions, a slow variation of one, or even two, units of reading resolution can be acceptable. It is believed, in fact, that the technical user has the competence, related to the statistical management of results, to be able to exploit the high potentials of the instrument without being fooled by the appearance. Different is the functioning condition of an instrument used in Legal Metrology, like the ones used in commercial exchanges. In that case is assumed that the users, both seller and buyer, are unskilled. Therefore the instrument has to give a clear result. To obtain such thing the builder rises artificially the unit of format (Fig. 3.3), getting rid in the result's presentation the less significant digit and, sometimes, even the previous one.

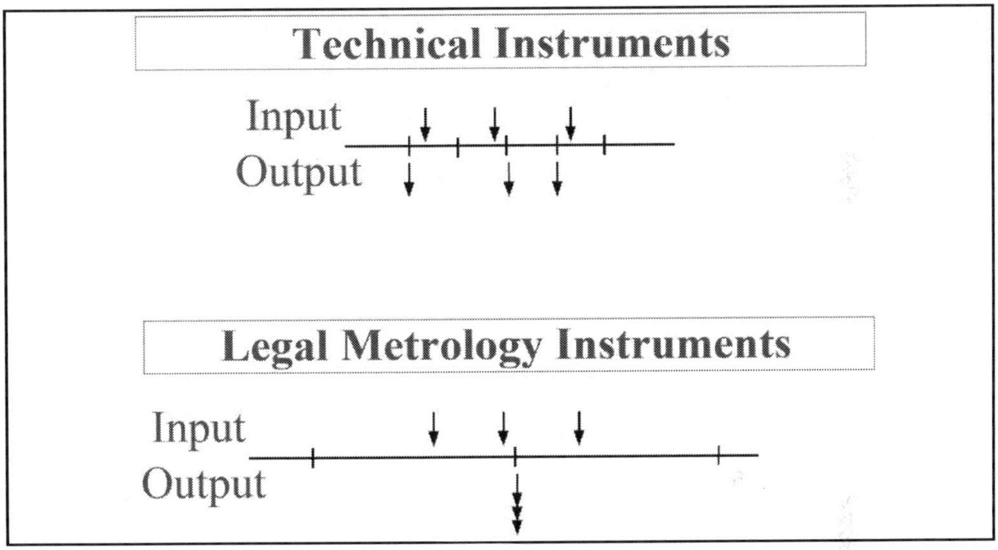

Fig. 3.3 the input values, scattered because of various causes of uncertainty, get transformed in the exit in the closest value to the unit of format. In case of use of technical instruments, the resolution is thin and the inputs will lead to different outputs. In case of legal metrology instruments, the resolution is artificially coarse. Frequently happens that all inputs for a measurand will lead to the same output. The figure shows how this can happen with the same dispersion of input data.

It can happen to see that two scales apparently equal, one set for technical metrology and the other for legal metrology, give completely different format indications. If, for example, the one used for technical metrology shows indications of a hundredth of gram, the one made for legal metrology applications gets modified to show indications of just a gram. In this way, with a large unit of format, the dispersion of the results, due to the various uncertainty factors, gets hidden and the results shown unambiguously.

What can be clearly seen for technical instruments is that the reading resolution is not an interval in which is reasonable to think that the measured quantity will be[7]. On the other side, a legal metrology instrument masks with the coarse resolution the effect of influence quantities and their modification of inputs; therefore precision cannot be detected with repeated measurements.

For this reason, it is a good practice to consider always reading resolution as contribute to the uncertainty. In case of technical instrument it is a small contribution to precision and, even taken it two times, one by itself and the second within precision, this will not affect significantly the uncertainty evaluation. In case we got data taken with a legal or legal-like instrument which cancel the effect of precision, the resolution contribute is one of the more important to be considered. As it is not always possible to know if data arrive from technical or legal type of instrument, it is better to consider the reading resolution contribute in any case. As we will see, for evaluating uncertainty is conventionally required the semi-interval of variability. We have to understand if the resolution is a total interval of variability and, in this case, its value shall be divided by 2. Let's consider the digital calliper described before with 0.01 mm resolution. If the measurand is nearly 2.02 mm, it will be possible to read, for instance, the indications 2.02, 2.03 or 2.04; even if the dimension of the measured piece is between two of these three values, we will read one or the other, nothing intermediate. We can observe that the indication nominally switches from 2.02 to 2.03 when the calliper sees 2.025 mm and from 2.03 to 2.04 when the calliper sees 2.035. Therefore when the output is 2.03 means that the measurand is seen to be between 2.025 mm and 2.035 mm, that is (2.04 ± 0.005) mm and this involved that the resolution is the total interval of variability, while the semi-interval is 0.005.

It is also evident in fig. 3.3 the reason for having a high resolution for the technical instruments. This is advantageous because they are meant to be used by expert operators; even if the indication is not exact, the normal management of the results allows to obtain better information than using an artificially lowered resolution. A digital calliper with a resolution of 0.01 mm, for example, if correctly used, gives indications that do not go further than ±0.02 mm or ±0.03 mm to the value of the measured quantity, giving a better indication than the one obtainable if the resolution should be lowered of 0.1 mm. *Reading resolution* of a measuring instrument is not always the unit of format. While, in case of normal use of the numeric scales[8], reading resolution corresponds frequently to the unit of format, for the analogue scale, as we will see, is always smaller.

Now, going deeply on the way to quantify reading resolution looking at the instrument scale, we shall consider separately the digital and analogue scales.

---

[7] frequently the users have not clear the difference from reading resolution, resolution and uncertainty and consider that an instrument with the reading resolution of 1 µm should give outputs with the uncertainty of 1 µm

[8] As we will see, for peculiar operations, the resolution of a numeric scale can be improved.

3.2.1.1 Digital scales resolution

For a digital instrument, it is usually simple to individuate the smallest increase of measured quantity that is possible to read. It is only necessary to observe the successions of the number presented. The most common situation is given in table 3.2 for instrument A: the increments of reading are constant throughout all the range. The value of those increments is the reading resolution, equal to 1 for instrument A.

The same happens for instrument B but, in this case, the last digit is fixed to zero; therefore the constant increment is 20, which is the reading resolution.

Instrument C is not at all common. Its aim is to give a sequence containing 5; even if the instrument allow the resolution better than 5, about 2, it has not the digit to present 2.5, therefore gives a step of 2 and a step of 3. In this case sometimes we can get the increment of 2, but what is always available is an increment better than 3, therefore, the resolution shall be considered equal to 3.

Instrument D is very common. As said before, the ratio of the range to the resolution is strictly connected with the real capacity of the instrument. Therefore to have a wider range, most common for the electrical quantities, the solution is to have on the same instrument some different ranges, keeping fixed the ratio with the resolution, e.g. one scale with range 1000 and resolution 1, one scale with range 2000 and resolution 2, one scale of 5000 and resolution 5. With modern electronic instrument the switching of the resolution can be done on the same scale; therefore we can have a range of 5000, but the increment is 1 up to 1000, then it switches automatically to two up to 2000 and, thereafter, to 5 up to 5000. If we describe the whole scale, we have to declare a resolution of 5, as this is the minimum increment guaranteed all over the range. Certainly designing our test and examining the data obtained, if we know that no measurement is higher than 2000, the resolution to be considered is 2.

Tab. 3.2 Examples of succession of indications of four types of measuring instrument

	Instrument A	**Instrument B**	**Instrument C**	**Instrument D**
First indication	121	1200	120	999
Second indication	122	1220	122	1000
Third indication	123	1240	125	1002

A deeper analysis will show two different resolution conditions of a digital scale. The one now described corresponds to the unit of format: if resolution is 1 we read 0. 1. 2. 3, ... and, if we read 2, our measurand can be nominally anywhere between 1.5 and 2.5. But in the moment the reading is switching between 2 and 3, we know that it should be near 2.5. This second resolution is in the order of a tenth of the unit of format, but is important to notice, that, in general, it can be used only with special procedures.

Let's quote two cases that apply effectively this special methodology. We can examine, as first case, the modality of determination of the drift of an instrument. The drift is a tendency of the readings versus time: a continuous increasing or decreasing regular change. Drift is common during the warm up phase of an electronic instrument (Fig. 3.4). It can be observed switching on the instrument, setting it rapidly to zero and evaluating the output variation in function of time. The output will be for some seconds on the zero, then, eventually, after a brief phase of alternation, it will commute to the first increment, let say 1, where it will stay again for some seconds and then will commute to the second increment after a brief phase and so on. It will keep a monotonic increasing or decreasing trend that has the characteristic shape of an asymptotic approximation to the value of stabilization. Clearly it is impossible to evaluate definitely the time interval in which the thermal drift move the output to 1, but the 1 indication will remain for the entire output interval 0.5-1.5, quite long, and on 2 it even longer, so that the variable time is not adequately defined. The commutation moment from 0 to 1, instead, that correspond with a good approximation to the value of 0.5, is neatly better defined in time. Therefore it is convenient to use it to define the observed phenomenon, as represented in Fig. 3.4.

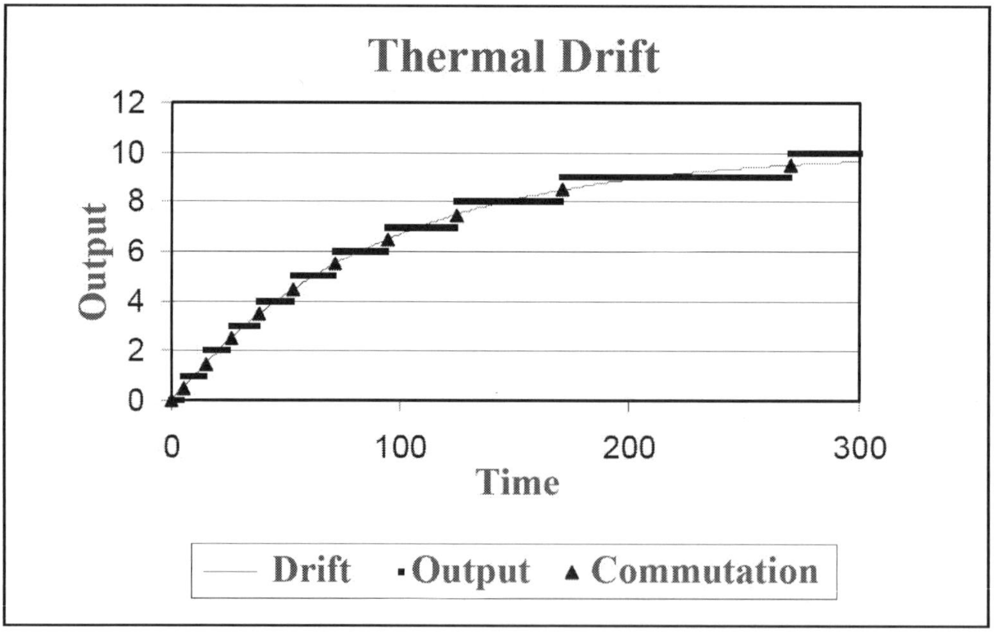

Fig. 3.4 Determination of the phenomenon of thermal drift. It is important to take note of how the drift trend is better defined to the commutation condition (little triangle), than the normal reading condition (continuous line).

As second case, the calibration of instruments with low resolution is reported. It can be adopted every time the reference has a higher resolution; therefore it is widely used in the sector of legal metrology. To improve readability it exploits the fact, mentioned above, that the switching condition has a width that is about one tenth of the reading resolution. Let's consider a balance having a unitary resolution, where every commuting will happen in the interval of a tenth (±0.05) around the values of 0.5, 1.5, 2.5, etc. Therefore, for example, the digital scales calibration can be done on the commutation points, using 10 masses of the value of a tenth of the resolution.

Let's examine a balance with 3 kg of maximum load (Tab. 3.3), with 1 g of resolution, which is calibrated following OIML Standards with sample masses of 600 g, 1200 g, 1800 g, 2400 g and 3000 g.

Tab. 3.3 Calibration table of a metric balance. Indication in grams.

Mass	**Nominal commutation**	**Real commutation**				**Mean error**	**Repeatability Range**
		1°cycle	**2°cycle**	**3°cycle**	**Average**		
0	0.0	0.0	0.0	0.0	0.00	0.00	0.0
600	600,5	600,3	600,4	600,3	600,33	0,17	0,1
1200	1200,5	1200,2	1200,4	1200,3	1200,30	0,20	0,2
1800	1800,5	1800,2	1800,3	1800,5	1800,33	0,17	0,3
2400	2400,5	2400,4	2400,6	2400,3	2400,43	0,07	0,3
3000	3000,5	3000,4	3000,7	3000,5	3000,53	-0,03	0,3

Let's analyse the first point of calibration as example of all the other consecutive points. We put 600 grams on the balance and we read 600. Thereafter we add, one at a time, some masses of 0.1 g, until the commutation to 601 happens. If, for instance, the indication commutes to 601 when a mass of 600.3 g is applied, we know that the balance measures 0.2 g more, because it anticipated 0.3 g before the commutation condition, which should be at 600.5 g. So it is possible to individuate the deviations from the ideal characteristic in a much better way than only reading the output indication.

3.2.1.2 Analogue scales resolution

Analogue scales present the measured information on a continuous variable, generally the angle of a circular scale or the length of a rectilinear scale. Nevertheless, the indication reading of the scale is discontinuous, for the limited eye optical resolution, which is about 0.1 mm in case of an observer without visual problems. It is immediately evident the problem of a mismatch between the real possibility of resolution of the eye and that to trace a scale with intervals[9] so small; therefore the scale division is always larger. What happen with the analogue scale of a watch that has a division each five minutes (fig. 3.5), apply to all analogue scales. Reading that watch no one will read 10:15, the nearest trait of the scale to the pointer, but 19:16 or 10:17. But how much have we to subdivide the scale division? The answer can be given taking into account that resolution is a capability of the instrument and reading resolution is not free; therefore information of the value of acceptable reading resolution shall be given to the user.

Fig. 3.5 A concrete example of reading an analogue scale. No one would read the time and say: "it is 10:15". Probably almost every one would say: "it is 10:17". It appears natural to divide the units of format of five minutes in smaller parts. Doing this the given information is improved

The adopted solution is conventional. The reading resolution can be $^1/_2$, $^1/_5$ or $^1/_{10}$ of the scale division. To allow to do the subdivision in 2, 5 or 10 parts correctly and to inform the user on which subdivision corresponds to the real resolution of the instrument, the geometry of the pointer and of the traits is conventionally defined: the pointer and the traits shall have equal and constant thickness (Fig. 3.6), which can be respectively $^1/_2$, $^1/_5$ or $^1/_{10}$ of the scale interval. Additionally, to consider the eye optical resolution, if the scale interval has a mean length (also in case it is an angle) lower than 1.5 mm, cannot be subdivided in 10 without using a magnifying glass. What said is summarized in Fig. 3.6

This process of subdivision of the graduation of an analogue scale is as simple as intuitive that operators prefer analogic scales because, if we have to work for hours, is less tiring than reading a digital scale. Analogue scales have also other advantages: we can see the speed of changed of the measured quantity, therefore if an instrument is used for control purposes, e.g. for reaching an established value of the measured quantity, the use of analogue instruments is important. Moreover, the reading on an analogue scale is quicker: try reading the time on analogue and digital watches. For these reasons on aircraft cockpits the main instruments are analogue, frequently duplicated with digital ones because these last usually have a better resolution.

---

[9] To be clear we use here the wording "scale division" or "unit of format" to indicate the smaller value of the quantity indicated by the scale, while with "scale interval" the angle or the length between the traits delimiting the scale division. This is not a general convention, but it can be useful here to be understood.

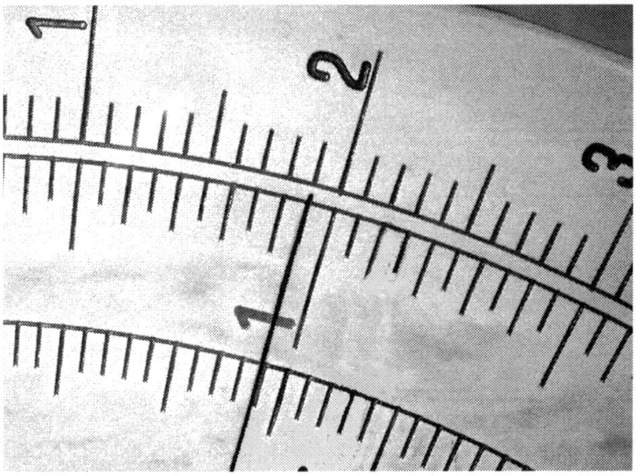

Fig. 3.6 Example of analogue scale in which we can see how the thickness of the index is constant and sensibly equal to the one of the traits delimiting the divisions.

Let's see in practice how subdivide the division of an analogue scale in $^1/_2$, $^1/_5$ or $^1/_{10}$. As said, good analogue scales have the thickness t of the index constant and equal to the thickness of the traits (Fig. 3.6 e 3.7) and intervals are n equal 2, 5, or 10 times the thickness t. The difference of these conditions is evident so that just looking at the scale we know the subdivision we can apply; therefore, from the scale division, dividing by n, we get the reading resolution. Remember that in case the scale interval is smaller than 1.5 mm, maximum subdivision is 5; if we want to increase the resolution, we have to use an optical device (sometimes instruments have installed a magnifying glass, or even complex optical systems).

After deciding in how many parts it is necessary to subdivide the elementary division, we have to evaluate how many of these parts are contained in the scale division of the pointer. This is usually done in a glance by expert operators, but even untrained people can give a correct answer. We can try on fig. 3.8 and 3.9. An untrained person gives answer within an interval of two units of resolution.

The complete reading must be done as follows:

a) evaluate the value of the principal division d, corresponding to the scale numbering, In the example numbering are 20 and 40, therefore $d = 20$;
b) count how many divisions are within a numbered division, in the example 10. Evaluate the value e of the division $e = d/10 = 2$;
c) evaluate the ratio between the scale interval and the pointer thickness, in the example $d = 10$. Evaluate the reading resolution $r = e/n = 0.2$;
d) read the value of the trait immediately before the pointer. In the example we have the numbered trait 20, plus 3 divisions: gives $20+3\cdot 2=26$

Guess how many resolution intervals are before the pointer, in the example 4, add 4 times the value of resolution to get $26+4\cdot 0.2=26.8$

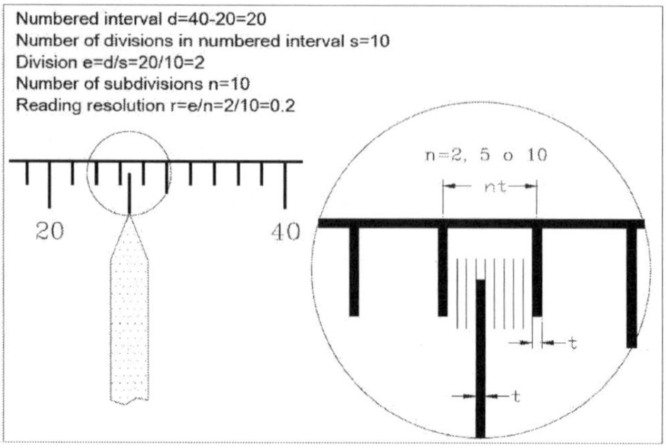

Fig. 3.7 Scheme to be adopted to read an analogue scale, determining the value of the unit of format, the subdivision n and the resolution r.

Sometimes instruments have both scales to allow getting all the advantage avoiding drawbacks. Fig. 3.10 represents the analogue and digital scale of a micrometer. The analogue scale is in two parts: a rectilinear part with a division of 0.5 mm (one trait in the upper part, the following on the lower part), and a circular part which division is 0.01 mm. We can try to read the analogue scale and compare the result with the digital scale.

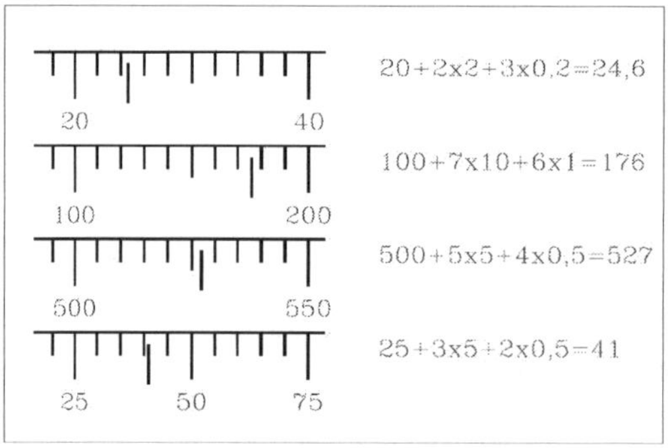

Fig. 3.8

Examples of reading on analogue scales

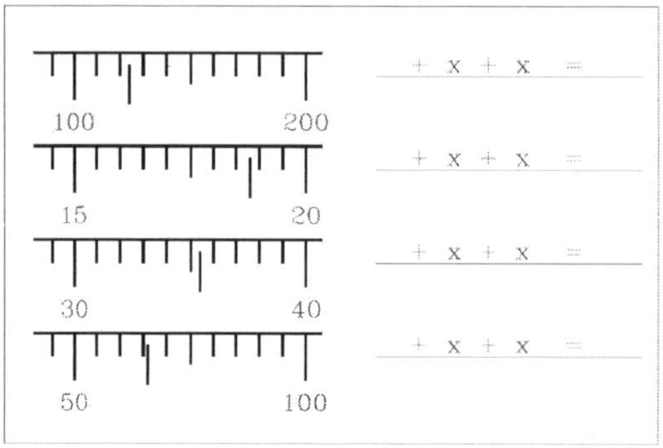

Fig. 3.9

Exercises of reading on analogue scales

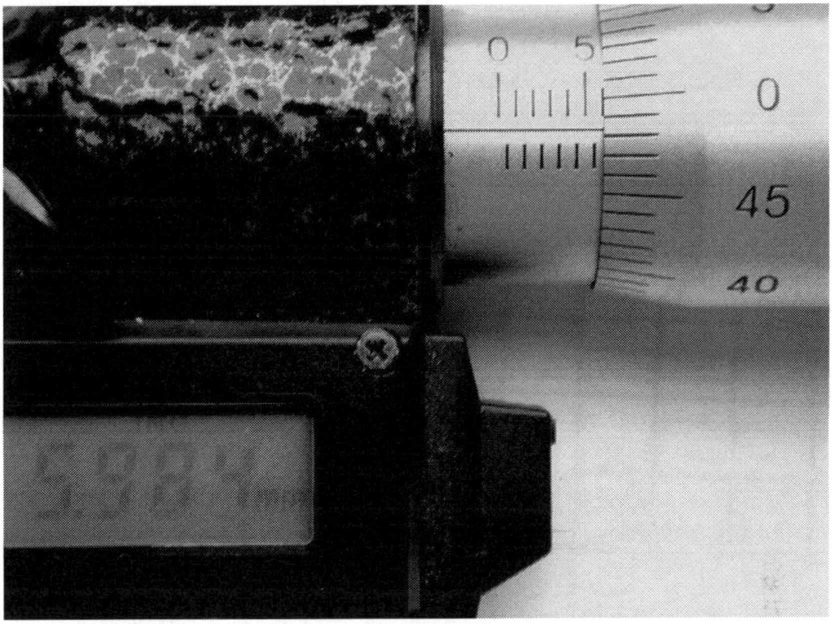

Fig. 3.10 Reading examples of an analogue scale of a micrometer. It is possible to check the reading correctness on the digital scale

Analogue scales can adopt an artifice for increasing resolution, invented in its modern form in 1631 by the French mathematician Pierre Vernier, therefore called vernier itself.

In Fig. 3.11 and 3.12 it is shown a short scale facing the main one. Its working principle is to have scale intervals shorter than that of the main scale; n intervals of the vernier have the length of n -1 intervals of the main scale.

Therefore each vernier interval is $1/n$ shorter than the scale interval (in fig 3.11 n=50).

When the zero does not arrive exactly on a trait of the main scale, as shown in fig. 3.12, the difference X shall be evaluated.

If k intervals of the vernier plus X cover the same length of k intervals of the main scale, considering d as the length of the main scale interval, we have:

$$k \cdot d \cdot \left(1 - \frac{1}{n}\right) + X = k \cdot d$$

$$X = k \frac{d}{n}$$

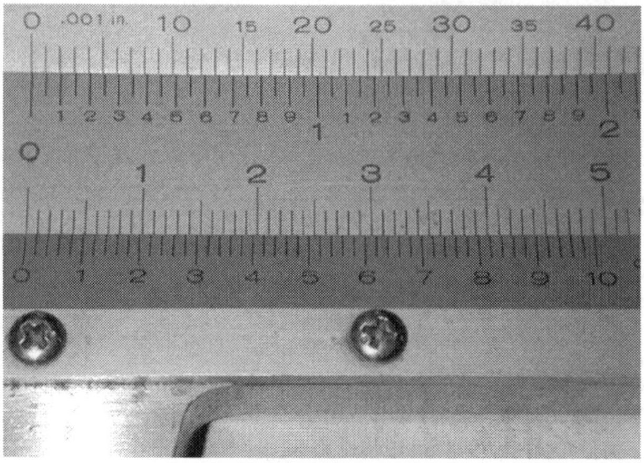

Fig. 3.11 *Vernier of a calliper. The vernier (see the lower scale), is composed by 50 intervals but it is only 49 mm long, therefore, every interval is 0.02 mm shorter that the interval long 1 mm of the main scale.*

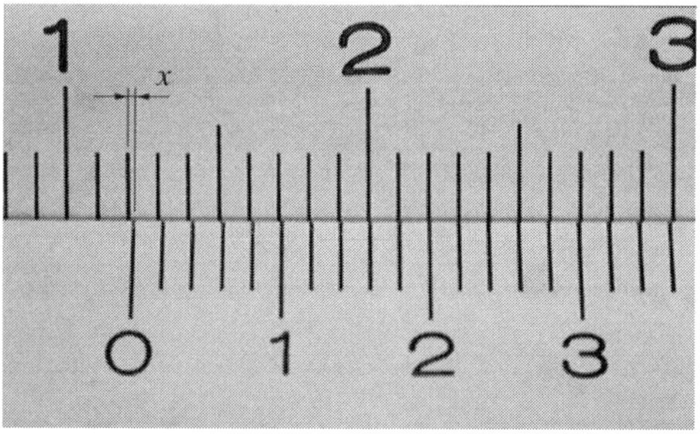

Fig. 3.12 *Reading of the vernier looking for its trait collimated with a trait of the main scale. In this figure it seems to be the trait after ten intervals (numbered 2) but looking carefully also the trait after interval eleven has a good collimation (looking very carefully the tenth is a little after and the eleventh a little before); therefore one can take the reading in the middle.*

In fig. 3.12 it seems that $k = 10$ and, being $d = 1$ mm, we have $X = 10 \cdot 1/50$, that is 0.20 mm (note that we have to write 0.20 and not 0.2, because with the vernier the reading uncertainty is on the second decimal). But looking carefully on the figure, we can see that also $k = 11$ could be accepted, therefore we can stay in the middle and accepting the result $X = 0.21$ mm, with a reading resolution of 0.01 mm.

Using the vernier, it is not necessary to do all the calculations reported above. It is enough to use the numbers engraved on the vernier scale. In our case at the tenth trait is engraved 2, which means 0.2 mm.

3.2.2 General introduction to other metrological characteristics

Before talking specifically about the other metrological characteristics, is important to find out which situations can be present in all measurement activities. We have to understand what happens when we carry out a measurement. Only then it will not be hard to assign, to the various factors, the connotations of the theoretical concepts defined in metrology. It is possible to acquire experience dedicating an adequate amount of time, but it does not require indispensably the use of sophisticated measuring instruments. Using a couple of common millimetre rulers (school type, possibly the first having a good quality and the second a low one), working in two groups with the same rulers to compare results, can let we know the meaning of the main metrological characteristics; but the stability cannot be evaluated with so simple measurement tools and trueness cannot the evaluated because it requires comparison with reference samples.

Practice

Measure with a ruler the lengths of the segments reported in Fig. 3.13, interpolating the reading at 0.1 mm, if we are using a good quality ruler, or at 0.2 mm, if we are using the other one. To determine the printing scale factor, use values listed for the first two segments.

Compare the measurement's results with the ones obtained by the other operators, trying to understand what can be the causes of the differences. They may depend on the use of different rulers (same operator, different rulers), or to the way different operator used the same ruler. Do not forget that it is natural, when we push to the limit the resolution of a measuring instrument, to observe differences in the results even strictly controlling the procedure (same operator, same ruler, repeated measurements in a small time interval).

Try to observe the influence of the time repeating our measurement after few days (it is unlikely that a ruler can change among time, but it is possible that from a day to another an inexpert operator could change the measuring procedure).

These first experiences should highlight how a measurement result is always uncertain. They show that small effects are masked by the natural dispersion of the results. To highlight the effects of the single factors (rulers, operators, instability differences) it is necessary to go over the problem generated by the results' dispersion. It can be done using the arithmetic average method. Try to evaluate the differences comparing single results and average ones. Try to get conclusions about the different validity of the information that can be obtained comparing these results.

Fig. 3.13 Length measure of segments[10], to be done with a good quality millimeter ruler, defining the scale factor with the quoted segments.

This exercise should highlight:

1. The replication of measurements at a distance of time could lead to significant changes of the results (although, as mentioned above, a measurement process so simple and unsophisticated should not be subject to instability or drift). The characteristic of a measurement complex that give slightly different results at different times is called *stability*.
2. The replication of measurements done keeping the operative conditions as constant as possible (same ruler, same operator) in a small time interval highlight a minimum measurement dispersion due, in large part, to the instrument. The ability of a measurement complex to give, under strictly controlled conditions, results that are not very disperse, is called *repeatability*.
3. More concrete information on the real behaviour of the measurement complex can be obtained replicating the measurement in the real use conditions. An example of working conditions are the results obtained in different times by different operators with different instruments. The dispersion of results obtained is quite important to evaluate the uncertainty contributions bounded to the use conditions (instead of having the *minimum* dispersion given by repeatability, we have an information of the dispersion *normally present in practical use*).

[10] Answers: a) 143.6, b) 143.3, c) 143.5 d) 143.2 e) 143.7

The ability of a measurement complex to give results not very dispersed, under not strictly controlled conditions, is called *reproducibility*.

4. An instrument that have a wrong scale (i.e. a ruler with larger or shorter intervals respect their nominal value) produces, on the same measured quantity, measurement results with a nominally constant error. Even the most accurate test of repeatability and reproducibility cannot highlight the presence of such irregularities in the function of the measurement complexes. These can be individuated only with the comparison with another instrument. Such procedure, usually carried out with a standard instrument or with a higher metrological characteristics instrument, is defined *calibration*[11]. This implies the research of the differences between the reference standard and the average value of the results obtained with the measurement complex to be calibrated.

The ability of a measurement complex to give, in strictly controlled conditions, results on average slightly different compared to the reference values is called *trueness*.

5. Finally, the global characteristic that gives the information of the range around the measurement result in which is reasonable to think that the value of the measured quantity is, assumes the name of *uncertainty*.

The uncertainty value is an indispensable complement to make usable the information produced by the measurement as seen in chapter 1.

The uncertainty is a fundamental part of the information contained in the measurement. If it is not expressed the measurement itself loses meaning. It is important to remind that is almost never necessary to do a deep and complete estimation of the uncertainty value. In fact, it is expressly indicated that the uncertainty values shall be stated with one significant digit, or maximum two.

3.2.3 Stability

Considering a technical measuring instrument in which the resolution is forced to the maximum.

If we use an instrument that gives us a dimensional measurement with 0.01 mm of resolution, we will have more complete information than using one with just 0.1 mm. The resolution is often taken as index of goodness of measuring instruments, but it is important to pay attention that the produced uncertainty contributions, even just by the use of the instrument, will not make

[11] It is important to distinguish calibration from a simple comparison between average measurement results and reference samples. The second one is meant to modify the measurement complex to give results closer to the reference samples'. This group of operations is called adjustment (VIM 3.11).

vain the advantages of the high resolution. An easy experience, with very sensitive electronic instruments, is the observation of the stability of the last digit. Usually it does not remain constant, but it often oscillates, getting higher or lower of one or two resolution intervals (one is common enough, while two resolution intervals is less frequent). What it is important to notice now is how the information is strongly influenced by the fact that the indication obtained measuring a constant entity varies over the time (the output value is unstable). This gives us the necessity to take into account the capacity of an instrument to give a constant output over time, when the input is constant. Such characteristic is called "stability".

Let's examine another example. A balance behaviour is affected by an instability of the zero (the display indication even when no weight is applied) that varies over time. This instability is, therefore, an effect present on the whole scale, because the variation on the zero affects every measurement. It is easy to see the effect of the instability of the zero when the instrument is on the zero or has a constant input. It is impossible, when there is no stable reference. This consideration is fundamental because it is easy to notice how, in many cases, the instability of the zero is not important, because is possible to restore the zero condition right before the measurement (tare operation).

In other cases, instead, the stability of the zero is fundamental. For example, when a balance is used to acquire the weight of the grains contained in a silo; certainly it cannot be emptied to control the zero. Here we see how the characteristic of the stability of the zero is fundamental for those applications.

An example of zero error is given by a monotonic variation in output (it varies continuously in the same direction) due to elastic-plastic behaviours, dampers, electrical condensers or heat capacity. This last effect can be easily observed when we switch on the electronic measuring instruments, for which an initial stabilization period, often scheduled and declared in the instructions, allows the appropriate heating of all circuits. In conclusion, it should be noted that, as appears from the given examples, the characteristic of stability must be taken into account in different ways for different applications of measuring instruments.

It should also be mentioned the stability of sensitivity[12], which is not easy to control as the zero stability. In fact, in that case, one must have provided a stable reference, which shall produce an output about 80% of the range. To realize the difference, just think of a balance of high capacity, such as a weighbridge of 80000 kg. To check the stability of the zero, it is sufficient to observe the variation of output when there is no load applied. To control, instead, the stability of the sensitivity, a high amount of weights is needed, indicatively at least half of the flow rate. It is not necessary to use standard masses; just a load of dry, clean gravel is enough. But we must avoid those items subject to weight loss or hygroscopic, (i.e. the ones that change their

[12] It is also called span stability or, specially for electronic instruments, gain stability or stability of amplification.

weight with moisture). In any case, the stability control of the sensitivity is considerably more complex.

Often the problem of the verification of the sensitivity stability is solved providing a stable reference instrument. Micrometres, for example, are equipped with a reference cylinder of a defined length. Electronic instruments, which due to their complexity are liable to be unstable, are equipped with simple accessory circuits, therefore stable, that can periodically be compared. Even many modern instruments perform in a fully automatic manner this comparison and also proceed automatically to the adjusting of sensitivity, in case this was incorrect.

Let's see now if the only characteristics of resolution and stability allow a concrete utilization of a measuring instrument (that allow to provide an information[13], useful and adequate to the purposes). Consider, for example, a continuous heat treatment. Its proper operation is guaranteed when the temperature in the different furnaces, of quenching and tempering, is maintained constant. Thermometers that control such temperatures may not even have a temperature scale, but the mere indication of a zone of acceptance.

Note how important is this observation, also from the economic point of view, that allows, in many cases, to save unnecessary calibration costs.

In fact, the stability control is usually very simple[14] and it does not have expensive as a calibration. Note that, for continuous processes, it could be, in many cases, impossible[15].

3.2.4 Precision

In the example mentioned above of continuous plants, stability and resolution are the only important metrological characteristics.

This is due to the fact that in the plant is reached a proper operating condition, generally controlled with other instruments (continuing with the example of the system of continuous heat treatment, the control is done on the hardness of the material treated) and the only task of its own instruments is to testify that the monitored parameters remain constant.

However, already in the case of a not continuous plant type in which the measured parameters do not remain constant over time but are reached cyclically starting from other values, the knowledge of stability only is inadequate. The instrument does not remain fixed at a level of measurement but, for example, starts every time from zero to reach the measured value. The fact itself of moving to the measured value introduces errors, such errors of mobility, hysteresis, inversion; which usually produce a greater effect than the

---

[13] As we will see, this information cannot be called "measure" and cannot be assigned to it the unit of measurement. In the examined case the number itself is not important, but it is important only the amount of its variations (small) over time.

[14] We might say that there is no need for a verification of the instrument if the indication is maintained sufficiently constant for the intended purposes.

[15] To calibrate such instrument can be necessary to remove it from the plant. This often means to stop the production during the calibration.

instability. A test on repeated measurements can, therefore, inform us about the performances when the measurand is not constant in time; the ability of giving nearly equal results in repeated measurements is called *Precision, as defined by VIM*:

> *2.15 measurement precision*
>
> *closeness of agreement between indications or measured quantity values obtained by replicate measurements on the same or similar objects under specified conditions*
>
> *NOTE 1 Measurement precision is usually expressed numerically by measures of imprecision, such as standard deviation, variance, or coefficient of variation under the specified conditions of measurement.*
>
> *NOTE 2 The 'specified conditions' can be, for example, repeatability conditions of measurement, intermediate precision conditions of measurement, or reproducibility conditions of measurement (see ISO 5725-3:1994).*
>
> *NOTE 3 Measurement precision is used to define measurement repeatability, intermediate measurement precision, and measurement reproducibility.*

The definition underlines repeatability and reproducibility condition plus a possible intermediate condition. The reason is that it is important to know this characteristic referred to the instrument alone or the whole measurement complex (instrument, operators, environmental conditions and measurand).

We speak first of *repeatability*, that is a feature strongly linked to the measuring instrument[16]. We want to evaluate the dispersion of the results produced by the fact that, to obtain the measure, we run a cycle starting from zero up to the measured value. The dispersion of possible outcomes is evaluated replicating the measurements from zero to a final value, usually near to the full rate, of a measurand much more stable than the instrument under test.

A well represented example of repeatability and reproducibility that highlights the differences between these two characteristics, can be done describing the required tests for force transducers. These are instruments that measure forces. They are controlled with standard force machines that, for example, are mechanisms that apply a series of masses in succession to the force transducer, in order to apply it the correspondent force-weight series.

The repeatability, as we said, is evaluated from the dispersion of the results that are obtained by returning several times in the same measuring conditions. But, as we attempt to highlight the effect of the instrument alone, we try to achieve a measuring cycle that introduces the minimum disturbance as possible.

---

[16] In the case of repeatability, we try to highlight the effect of the Instrument making other contributions as small as possible.

For a force transducer the repeatability is, therefore, evaluated simply applying several times, slowly, the same group of masses to the force transducer. It is an operation made with the maximum care, with all the attention to not introduce disturbances, in a short time and with a single operator. This operating condition tends to represent an ideal condition and, therefore, provides as information a limiting value: in practice, the minimum dispersion of the results obtainable with that instrument. The advantage in the evaluation of repeatability is that, with a procedure which generally is the simplest and fastest for the kind of measurement examined, we know immediately the corresponding limit condition: in any case one cannot hope to obtain a lower dispersion.

The information on the repeatability does not completely solve our problem. It defines the characteristic of the instrument, but in reality we should always take into account the behaviour of the measurement complex as a product of the interaction between instrument, operator, environment and measured quantity and disturbances introduced into normal measurement.

In the case of the example of force transducers, the interaction is strong since they are tools that are often used in the field. They are transported to the company where measurements must be made, reassembled under the operating conditions, placed in the position in which they make the measurement. After that, they get dismantled, packaged for transport and transported again.

The result can consequently be affect by the mounting condition and the position as load receiver, which cannot be identical in different applications. For this reason, the relevant standard document[17] requires tests of *reproducibility*, which must represent more closely the real operating conditions and requests that, in the measurements replication, the force transducer is repositioned several times in the loading area of the standard force machine. Then it has to be rotated around its loading axis and, between two successive test runs, it gets, at least once, disassembled, packed as expected for the transport, reassembled and put back in the force standard machine.

This example clarifies well the difference between repeatability and reproducibility. Also allows us to highlight how the evaluation of repeatability is simple and fast, while the reproducibility is both longer and more complex. This characteristic of repeatability makes it very useful in operations of intermediate verification (ISO 17025), since it makes possible, with a limited cost, the execution of a check on the proper functioning of the measuring instrument.

We must, therefore, always pay attention to the real meaning of the data declared as repeatability or reproducibility. Knowledge of the characteristics of reproducibility allows us to use the instrument in many cases of practical interest. Note that for the conditions of repeatability and reproducibility it is not necessary to use a calibrated standard; a stable measured quantity is enough.

For example, the dimensional inspection of a piece can be done with a displacement transducer positioned on a support in order to give an indication 0

[17] ISO 376:2011-- Metallic materials -- Calibration of force-proving instruments used for the verification of uniaxial testing machines

when the reference piece is introduced, which, in practice, can be a piece of normal production already accepted by the customer.

Repeating the measurement of the same reference piece several times in different conditions,, each time as if it was a different piece inserted in the production process, it is possible to determine the reproducibility and, therefore, to evaluate whether the dispersion of the results (that are obtained in the replicated measurements) is small enough to be acceptable for the purposes of measurement[18].

Verified that the reproducibility is adequate, we can use the controlled instrument in comparison measures without it being necessary to submit it to a full calibration.

3.2.5 Measurement Trueness and Bias Error

In some cases, measurement operations turn to determine an actual measurement, assigning a number to the measured quantity accompanied by the unit of measure. In previous cases, concerning the determination of the proper operation of a continuous process or test for comparison of produced pieces, the measuring instrument might not even had a scale on which to read the numerical result, but only a highlighted area indicating the acceptability or not of the result.

If instead the information of measurement must be expressed by a number and its unit of measurement, it is necessary that it is equivalent to measurements of the same quantity eventually done by others. In previous cases, possible systematic errors constant of the instrument were not important because, in the example of a comparison between two pieces, the same systematic error would be present in both measurements. Therefore, the purpose of evaluating the difference between the two measured pieces would have been achieved successfully.

If, however, the manufacturer has to qualify his production, it needs a measure that must be free from systematic errors. In this case a complete measurement expressed by the value, the units and the uncertainty of measurement[19] is necessary.

It is no longer possible to act without an assurance that the adopted measurement process will provide, within the stated uncertainty, the same value obtainable with other instruments; these results have to be compatible.

---

[18] The control conditions are never absolute, but must be established on the basis of the results required. For example, if the required dimensional tolerances are ± 0.02 mm and the dispersion observed by replicating the measurements on a single piece is ± 0.01 mm, only ±0.01 mm would be left for the dimensional variability of the piece. This may lead to an excessive waste due to the testing process rather than to the production process.

[19] As we have seen, the uncertainty value is necessary to transfer any information, whether it is a question of complete measurement or even partial information regarding stability and comparison only. In the latter case, for example, the evaluation of the dispersion just linked to the reproducibility is an index of the uncertainty, for understanding what variations are normal for the measuring process, therefore, not attributable to differences between the measurands.

For this reason, quality standards require that any instrument used to provide measures has to be traceable to national standards and that each result is associated to an uncertainty value.

This helps to evaluate compatibility of results obtained with the national standard and, therefore, with a sort of transitive property, compatible with each other.

The characteristic obtained by the comparison between the results of measurement of an instrument with those the national standard is the *trueness*, as defined by VIM:

> *2.14 measurement trueness*
>
> *closeness of agreement between the average of an infinite number of replicate measured quantity values and a reference quantity value*
>
> *NOTE 1 Measurement trueness is not a quantity and thus cannot be expressed numerically, but measures for closeness of agreement are given in ISO 5725.*
>
> *NOTE 2 Measurement trueness is inversely related to systematic measurement error, but is not related to random measurement error. NOTE*
>
> *3 Measurement accuracy should not be used for 'measurement trueness' and vice versa.*

Note the fully theoretical definition, evidenced by the term "infinite". The problem is that practically we have always some random errors, but we have seen that the effect of random error is reduced if we take the average; theoretically, since the variance of single data being divided by the number of data averaged, if we average infinite data, we cancel the effect of random errors, but practically, few data are normally used (standards frequently ask to take the average of 3 or 5 data); therefore the value of the systematic error obtained has its random variability.

Generally trueness is indicated numerically by the amount of its contrary, the systematic error due to the difference from national standard, specifically called bias error or simply bias, as defined by VIM:

> *2.18 measurement bias*
>
> *estimate of a systematic measurement error*

3.2.6 Accuracy and Uncertainty

In most cases we are interested in the total variability of measurement results, being it due to random or systematic errors. Therefore we need a characteristic that combine these two types of effects, called accuracy as defined by VIM:

2.13 measurement accuracy

closeness of agreement between a measured quantity value and a true quantity value of a measurand

NOTE 1 The concept 'measurement accuracy' is not a quantity and is not given a numerical quantity value. A measurement is said to be more accurate when it offers a smaller measurement error.

NOTE 2 The term "measurement accuracy" should not be used for measurement trueness and the term measurement precision should not be used for 'measurement accuracy', which, however, is related to both these concepts.

As for many other characteristics, also accuracy gets a value from its opposite, the uncertainty, as defined by VIM:

2.26 measurement uncertainty

non-negative parameter characterizing the dispersion of the quantity values being attributed to a measurand, based on the information used

NOTE 1 Measurement uncertainty includes components arising from systematic effects, such as components associated with corrections and the assigned quantity values of measurement standards, as well as the definitional uncertainty. Sometimes estimated systematic effects are not corrected for but, instead, associated measurement uncertainty components are incorporated.

NOTE 2 The parameter may be, for example, a standard deviation called standard measurement uncertainty (or a specified multiple of it), or the half-width of an interval, having a stated coverage probability.

NOTE 3 Measurement uncertainty comprises, in general, many components. Some of these may be evaluated by Type A evaluation of measurement uncertainty from the statistical distribution of the quantity values from series of measurements and can be characterized by standard deviations. The other components, which may be evaluated by Type B evaluation of measurement uncertainty, can also be characterized by standard deviations, evaluated from probability density functions based on experience or other information.

NOTE 4 In general, for a given set of information, it is understood that the measurement uncertainty is associated with a stated quantity value attributed to the measurand. A modification of this value results in a modification of the associated uncertainty.

For example, a producer measures the average diameter of 5.003 mm of a cylindrical pin. He knows how the results of the measurements are dispersed. To avoid disputes, whenever he provides a measurement result he indicates also the uncertainty. If the data that he has provided is 5.003 mm ± 0.002 mm and a second or third party says that the diameter is 5.006 mm ± 0.002 mm, there is an interval of uncertainty overlap (in the specific case the interval between 5.004 mm and 5.005 mm), for which the two results are compatible each other. It cannot be said that are significantly different.

As we have seen, uncertainty, that puts together the information of reproducibility and trueness, is a very important characteristic and, for being estimated, it requires a wide use of statistical methods.

3.3 MOST COMMON MEASUREMENT'S ERRORS

3.3.1 Global schematization

The schematization of the characteristics of a measurement complex, especially for the part concerning reproducibility and trueness, frequently arises from common types of measurement errors. To understand it, we must think about the factors that affect a measurement result. Let's look at a very common case: the dimensional measurement of a mechanical piece. We consider now an influence factor present in many measurement processes: the temperature (considering, for simplicity that it only applied to the piece itself and not also to the measuring instrument). In such a case, we must consider that the change in temperature produces a dimensional variation of the object itself[20], described by the phenomenon of thermal expansion. To a temperature change ΔT of a body, it corresponds a contraction or dilation ΔL of its length. This relationship is, in first approximation, a proportionality law:

$$\Delta L = \alpha \Delta T$$

where the proportionality factor α is called *coefficient of thermal expansion* of the material of which the body is made.

To switch to the errors that can produce thermal expansion in dimensional measurement, we must note that there may be three different modalities of action of the same physical phenomenon, and emphasize straight away that the type of error is not necessarily linked to the physical phenomenon of origin, but to the way the error itself can be managed.

Let's examine these three different situations.

The first can be exemplified by imagining to operate in unconditioned environment, with large variations in temperature. It is likely, in this situation, that is possible to verify the mathematical relation between the length and

[20] For this reason the definition of length must be accompanied by a reference temperature, set at 20 °C by ISO 1:1974.

temperature changes. Being valid the mathematical relationship the values ΔL can be calculated and used for correcting the measurement result.

To describe the second situation let's imagine to operate in a conditioned environment.

Temperature variations are very small (for example, in metrological rooms of the order of a few tenths of degree Celsius), then also the corresponding length variations are very small, so small to be confused with those produced by other sources of error (such as the differences in temperature within the piece and from the room temperature, contact pressure of the probe, vibrations, the resolution of reading). We cannot verify the mathematical relationship between the change in length and the cause that produced it, because many causes can produce effects of the same size (so small that cannot be isolated).

The third situation is that of an automatic instrument operating in a thermostatic room.

These instruments, now very common, work stand-alone (we let them unattended for long periods of time while they do their job). If during one of these measurement periods a colleague open for a while the thermostatic chamber (often without even feeling the gravity of that action, so not reporting what happened to us), this would lead to a sudden variation of temperature.

Analysing the data, we would observe that, one or more results will be incomprehensibly different from the others.

The difference between these three situations is quite clear. In the first case the error, defined *systematic,* can be corrected but also other different methods can be applied to eliminate these effects:

- **correction,** which consists in the calculation of the error and its subtraction from the measured value. In the shown example, the temperature T is measured and the thermal expansion coefficient α of the material identified so that we can calculate the error $\Delta L = \alpha(T\text{-}20\ °\text{C})$ and correct the result;
- **compensation,** which consists in associating to the measurement a physical phenomenon that compensates for the variations of the result. In the shown example, the compensation can be obtained using an instrument with the scale of the same material of the measurand. In such a case, if the temperature change has a slow variation, it will vary nearly together the size of the object and the dimension of the scale of the same entity. Therefore, the size of the object will always be in the same graduation of the scale[21];

[21] This is the reason why it is possible to make dimensional measurements with good accuracy even in unconditioned environments. In many cases pieces of steel are measured with steel instruments, then there is a good compensation of the temperature effect. Nevertheless, we must keep in mind that this does not happen when the piece is made with a different material or when there are temperature differences between the piece and the instrument.

- **removing the cause**, this means to identify the cause of the error and to remove it. In the example of sensitivity to temperature variations, the elimination can be obtained avoiding the changes of temperature, e.g. operating in a thermostatic chamber.

In the second case, no relationship of cause - effect can be established. Therefore it is impossible to obtain any deterministic elimination of the error. However, it can be observed that the conditions of many causes of small errors vary in an arbitrary manner allowing a statistical treatment. Such errors are called *random errors*.

The third situation is described observing that there may be cases in which the error, although random and not connectable with its cause, at the same time it is not compatible with the random ones naturally present, to do statistics together.

These are cases in which we have real measurement accidents, statistically called outliers. Whenever we have an accident of this type, its causes remains unknown, and the errors cannot be deterministically corrected or managed with statistical processes.

We will see that they are managed by the principles of exclusion that are statistical methods based on the fact that accidents are rare events and that, therefore, identified as possible accidents (outliers) all the measurement results of such entity to be only slightly probable[22].

In conclusion, we can use the following global typology of errors:

- **systematic errors.** They are characterized by the identification or detectability of the relationship that connects them to the cause that produces them. This implies that their value is higher than the cloud of random errors, in which, otherwise, they would be confused. It is possible to correct, compensate or eliminate them, even if not completely;
- **random errors.** Are the remnants of systematic errors after their correction, compensation or elimination, and of the effects of other influences on the measurement (characterized by a random nature, which makes them undetectable in a deterministic manner). Their condition of randomness, together with the high number of factors that produce them and to the fact that they have very similar effects of entities, allows to manage them with the statistic of the normal distribution;
- **accidents of measurement.** Are characterized by being unpredictable, but quite rare and they are in higher amount compared to the cloud of normal random errors.

It is important do not confuse the concepts of error and uncertainty.

[22] The use of the principles of exclusion is not unanimously accepted. It is important to emphasize that, in any case, it should be used with great caution. While it is true that accidents are rare events, it is not true that all the rare events are accidents.

The uncertainty contributions are what remains of the errors after it was done everything possible to eliminate them, as shown in Fig. 3.14.

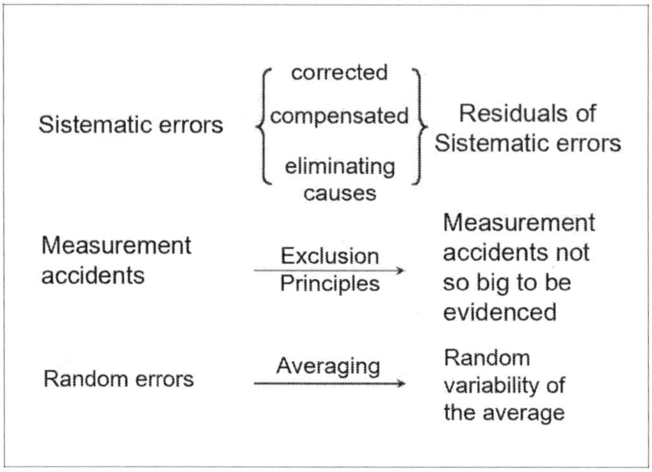

Fig. 3.14 Scheme of elimination of systematic errors and measurement accidents. What remains, along with the random errors, constitutes uncertainty.

3.3.2 The main specific measurement errors

Some kinds of measurement errors are very common, so it is appropriate to describe and discuss them specifically. For this purpose, we will use the schematic of Figure 3.15, which represents them according to their origin and the managing way.

For each type of error it is assigned the attribute of random or systematic, which must be understood as indicating only the most common situation, since, as we have seen, the fact of being random or systematic does not depend on the etiology of factor itself, but how it can be managed.

Cause	Measurement errors	Type
operator	Reading error	random
instrument	Mobility error and mobility threshold	random
instrument	Inversion error	Systematic
instrument	Linearity error	Systematic
instrument	Hysteresis error	Systematic
Operator-instrument	Zero error	Random- System.
Accommodation and environment	Error of the reference standard	Systematic
Accommodation and environment	Error due to influence quantities	Systematic
Process	Change of measurand due to measurement	Systematic
Dynamic	Time-frequency induced variability	Systematic

Fig. 3.15 Scheme of the main measurement errors

3.3.2.1 Reading error

The reading error was traditionally attributed to the operator, because with the use of instruments with analogue scales this was considered more common. For instruments with a numerical scale, instead, the reading error depends only on the resolution. Cannot be excluded also trivial, but not rare, transcription errors. So it is strongly recommended to use interfaces to connect instruments to the computer in order to read automatically the data and to be ready for processing while minimizing transfer errors.

In the case of the operator with analogue scales, the most common cause of reading error is the parallax error, due to the fact that only with special devices indexes can be made coplanar with the measuring scale. In general, they are separated from the scale by a certain distance, for which the condition of collimation depends on the direction in which the scale is observed, as shown in Figure 3.16. The error due to the direction of observation is, in fact, called parallax error.

To avoid parallax error, some methods, which tend to define the direction of observation, are adopted.

One of the adopted methods is the blade pointer, constituted by a thin plate (Fig. 3.16a). The development of the plate in orthogonal direction respect the scale is such that, if observed in an inclined direction, it shows part of one of the sides, while if observed in the right direction, it just shows its thickness (much smaller than one of the sides). The direction in which the plate looks narrower is, therefore, the one of the correct reading. This is highlighted in the images represented in Fig. 3.17.

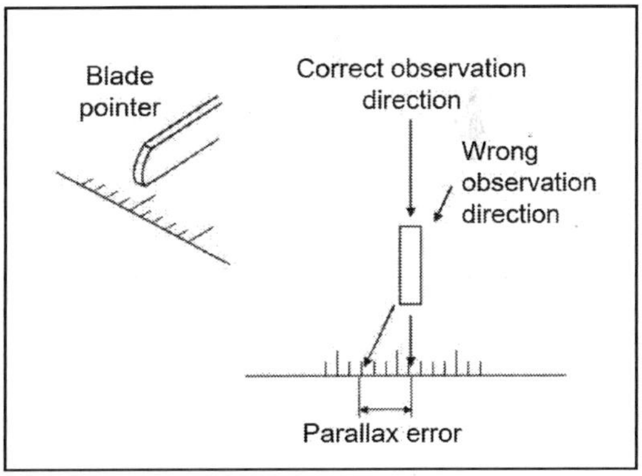

Fig. 3.16a Parallax error happens when the observation direction is not orthogonal to the plane of the scale. The blade pointer makes such error possible to be avoided, because the correct direction is evidenced by the pointer itself, which does not show its sides and appears as narrow as possible in the right observation direction.

A second method to avoid the parallax error is the use of a reflective surface coinciding with the plane of the scale (Fig. 3.16b). In this case, the direction between index and its image drive the observation direction. If the observed direction is wrong, we will see the index side by side with its image, as represented in Fig. 3.18, while the correct condition is when the index covers its image, making it impossible to be seen.

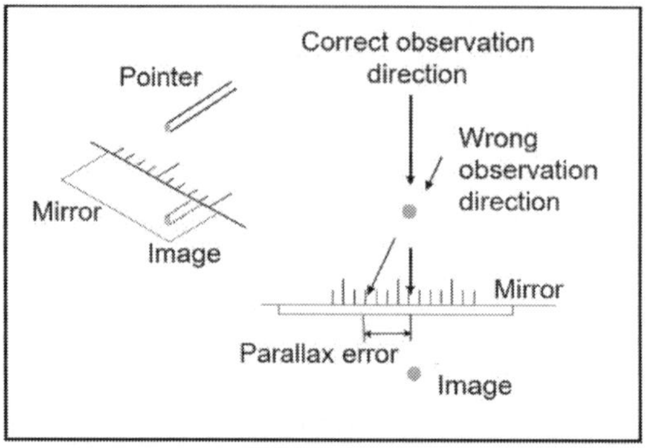

Fig. 3.16b To avoid parallax error is common to use a mirror on the measuring scale plane. The correct collimation is indicated by the overlap condition of the index and its image.

a)

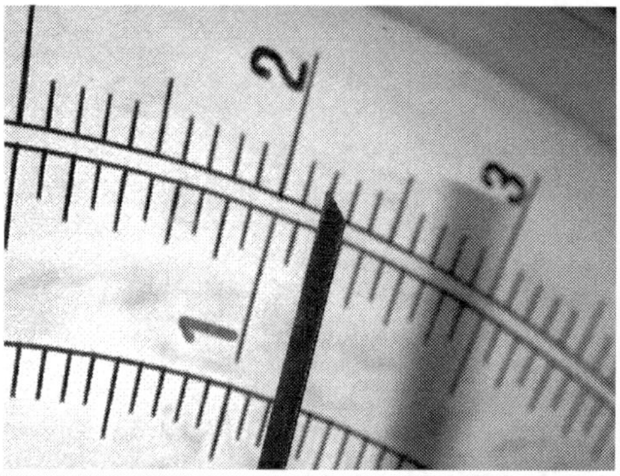

b)

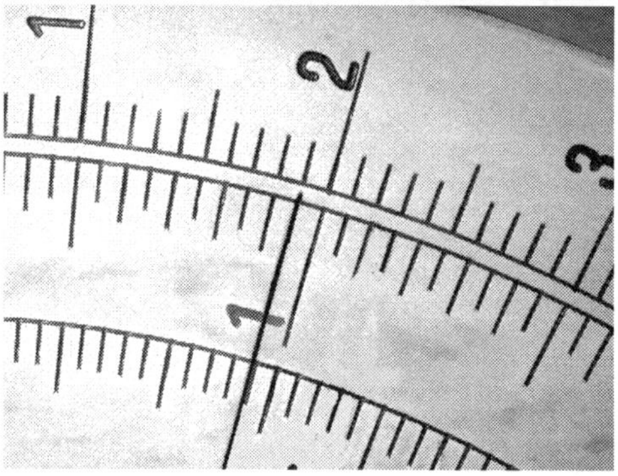

Fig. 3.17 The apparent thickness of the index is neatly higher in the condition (a) in which it is observed in the wrong direction, respect the right condition (b)

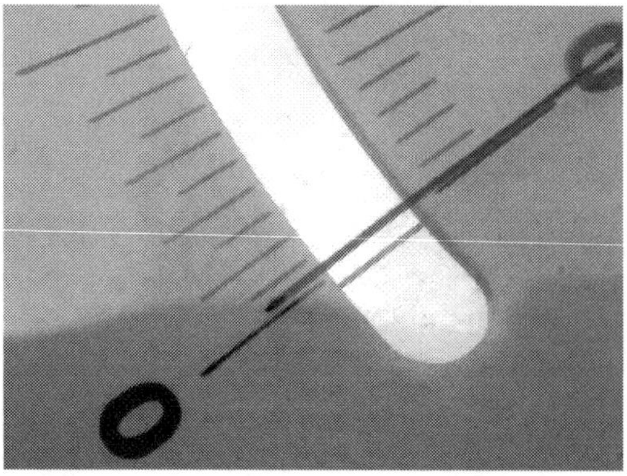

Fig. 3.18 Artifice of mirror scale. The presence of the parallax error is indicated by the separation between the index and its image.

Both resolution and parallax errors are caused by random factors, which make them random. It is important to notice that sometimes the distance between the index and the scale is not so evident, for example, in case of a micrometer described above.

As we can see in Fig. 3.19, such distance exist and can be cause of not negligible parallax errors. Besides this, often there are no reading expedients that help avoiding such errors; therefore, it is really important to be really careful to get the correct viewing direction.

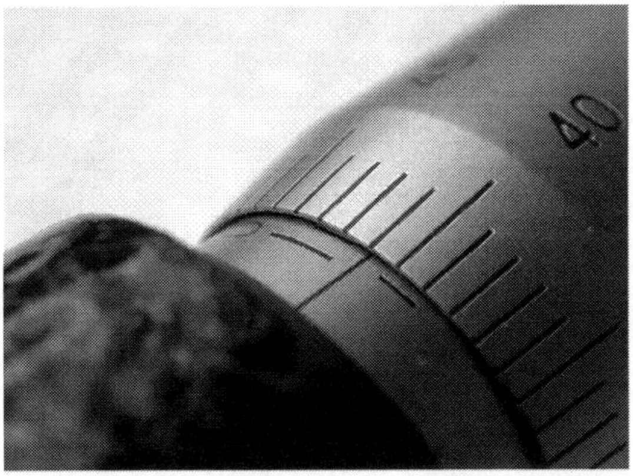

Fig. 3.19 Even when not evident, the index can be distant from the scale surface producing a parallax error.

Besides the case about the reading of the index position on a measurement scale, similar problems arise in all those cases in which we must do the collimation of reference lines with points of an observed object.

This situation was certainly seen by everyone doing measurements with flexible meters or rulers, which require before to collimate the zero with the initial point of the measurand, then to identify the scale division coinciding with its final point.

We are in the same situation when observing the position of the free surface of the manometer liquid in a column pressure gauge, or when we want to lead to collimate, with a desired scale division, the level of the liquid which is poured into a graduated cylinder. This is a situation which is often present in the use of measuring microscopes, cathetometers or telescopes, in which the measuring scale is optically superimposed on the object to be measured. Even in this case the most common error is the parallax one, but, collimating an object, a psychological habit of the operator can produce a systematic error. It may be instructive to examine a Japanese experience in the case of measurements of Vickers hardness indentations[23]. They appear at the microscope of an almost square shape on which we must measure the length of the diagonals. The measurement is made by aligning the diagonal with a reference line, thereafter bringing the two moving lines, orthogonal to it, to the

---

[23] Hida N., Yamamoto K., On the determination of absolute value of diagonal length of Vickers indentation. Bulletin of the National Research Laboratory of Metrology No 25, 1970

two opposite vertices of the indentation. Japanese have run numerous tests by six different operators: three experts and three novices, and have noticed that the measurement differences between the operators were systematic (an operator always tended to give a greater dimension and another a smaller one). The reason was because the real vertex is a point of infinitesimal dimension, therefore it is not shown by the microscope: the image of the indentation ends a little before to arrive to the vertex. Now some operators try to put the collimating line at the last point shown by the microscope, while others try to figure where the visible sides of the square would intersect. Certainly the last ones measure more correctly, but always longer than the first ones. This has a character that transcends the specific hardness measures, because it highlights how the collimation operations are done according to a mental model systematically different from person to person.

It is always recommended to remember that when we do a collimation, it brings into discussion the visual resolution. This must be taken into account, as shown for the reading of the analogue scales, even when we do collimation operations. Obviously, if the object to collimate is not well defined, we have to assume that the resolution of sight will be worse.

3.3.2.2 Mobility error and mobility threshold

These concepts are closely related to the resolution one: the determination of the minimum measurable increment. The reading resolution represents that, regarding the reading scale. A well-built instrument, functioning correctly, should have such value as real limit. However, if poor maintenance let increase friction significantly, or if the manufacturer has adopted a scale with a finer resolution than acceptable, the minimum measurable increment may be higher.

To understand better these characteristics, we will make the example of a balance verification, instrument for which the determination of the *mobility threshold* and *mobility error* is relatively easy. Furthermore, it is an instrument at least in part mechanical, for which the concept of static friction is an important help to understand the operation.

We begin to describe the concept of a threshold of mobility. Imagine a yoke of a traditional equal arm balance with its knife suspensions. The application of a weight on one of the plates produces the rotation of the knifes on their support, but such rotation is opposed by the friction and, in particular, if the knifes are stationary, by the stiction. Therefore, if we apply a very small weight, knifes and the index connected to them, will not move. Then, we will slowly increase the weight until the force produced overcomes the stiction up to the incipient movement. The amount of weight that creates the movement is the mobility threshold. For balances, the test mode is very simple: we use a series of equal weights, very small, of the order of one tenth of the reading resolution, and we begin to add them on a plate (with many precautions in order to avoid giving pushes or dynamic forces while they are being supported). It is noted several times (Fig. 3.20), at different points of the scale, which is the weight increment that produces a variation of the indication.

This testing method provides an indication of the operating conditions and it is valuable in the periodic inspections because, with a very simple operation, it would be immediately evidence any functional problem. However, it does not represent the normal condition of use, in which, instead, a certain weight is leaned in a condition of the scale normally quite distant from the equilibrium point. The yoke, therefore, arises in motion, reaches the point of balance and exceeds it by inertia, then it goes back in a series of damped oscillations to a stop.

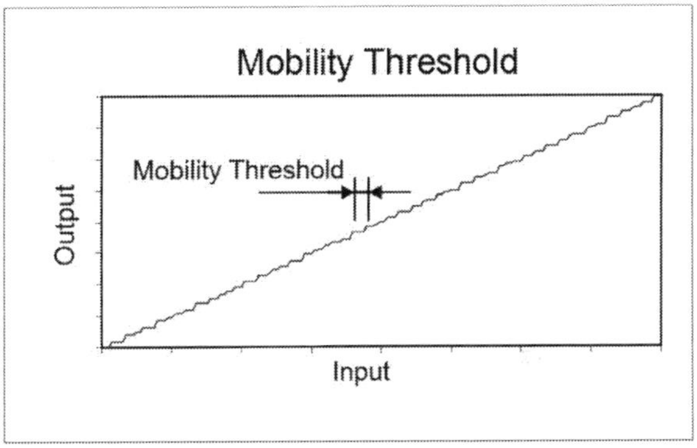

Fig. 3.20
Diagram of a test of threshold of mobility determination. The mobility threshold shall be taken on the input.

But it will be able to stop, in a completely random way, with a final ascending or descending swing phase. Therefore, also in this case, the friction will have its role and, in order to evaluate the effect, we apply a load X consisting of a main value plus a little overload ΔX (Fig. 3.21) of the order of ten to thirty times the resolution.

From that starting point the overload ΔX is removed and we add it again slowly, in order to be sure to reach the equilibrium point for the load $X+\Delta X$ with a growing movement. Then we record the output A in this condition. We now add another overload ΔX and then we remove it slowly, in order to reach again the balance point for the load $X + \Delta X$ this time with descending motion, and we record the output B. The friction will always act opposing to the movement, therefore, the movement in the increasing direction the friction will produce an output A lower than the expected one. In the case of decreasing loads, the friction will produce an output B higher than the expected one. The difference B-A will be an indication of the double effect of friction (twice the mobility error). In practice, the limits of readability of the results are not only related to the reading resolution, but, for evaluating the actual resolution, we must determine the mobility threshold: the first, in case the equilibrium point is achieved sufficiently slowly to ensure the monotonic increasing condition, the second, in case it is reached with oscillation.

3.3.2.3 Backlash error

Always taking into account mechanical instruments, it is not difficult to understand the meaning of the backlash error, which is linked to the presence of clearance in the kinematic chain of the instrument. If the direction of variation of the input is changed, a first part of the return of this will be used to recover the clearance.

Only when it will be totally recovered the output will start to decrease, as described in Fig. 3.22.

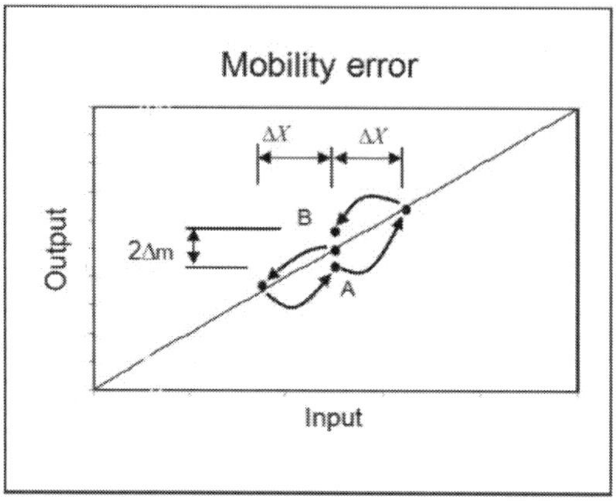

Fig. 3.21 Scheme of the mobility error Δ_m. Test

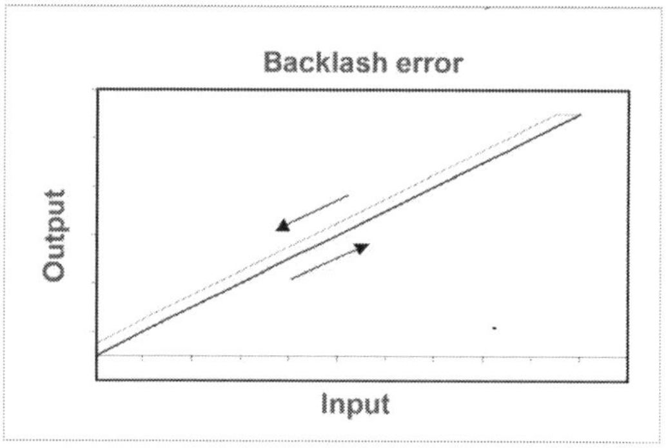

Fig. 3.22 Scheme of the behaviour of the backlash error, which has the characteristic of being substantially constant over the whole scale

The output will be different for the same input value, depending on whether this value is reached for ascending or descending values. This phenomenon is known by operators that use mechanical instruments or machine tools, whom, to avoid their effects, pay great attention to reach the measurement condition always from the same direction (ascending or descending). This explains what we see often. The operator deliberately exceeds the measured value assigned and then he reaches it coming back.

3.3.2.4 Linearity error

Apart from a few instruments that have deliberately a parabolic, logarithmic or other characteristic curves, in most of the instruments we look for a condition of proportionality between the input and output. This implies that the characteristic line, which describes the relationship between input and output, is straight. In practice, no characteristic is perfectly linear: small deviations constitute the linearity errors.

It must be said, in general, that the linearity error is due to the deviations of the real behaviour of the instrument in relation to its ideal characteristic curve, not only in the case of straight characteristic, but also when it is parabolic or logarithmic. The nonlinearity does not consider so much the irregular deviations due to random factors, but the fact that the systematic behaviour of the instrument is different from its ideal characteristic.

Let us examine, for example, a column shape load cell: this instrument, based on the Hooke's Law, consisting in a column with a constant section of elastic material, for example steel, which is loaded in the axial direction by the force *F* we want to measure. Hooke's Law says that the axial deformation[24] is proportional to the applied force *F*. The coefficient of proportionality depends on the area of the resistant section.

A column with a large cross section will deform less than a column of small section. It can be seen that the deformation is inversely proportional to the area *A* of the section. Therefore, the conventional deformation is described by the formula:

$$\varepsilon = \frac{F}{EA}$$

where *E* is the elastic module of the material, with a value as greater as the material is rigid. As it can be seen, the relationship between force and deformation is linear, and then the load cell should have a rectilinear characteristic. However, as we can see deforming any material flexible enough to present a visible deformation, the effect of the axial load is not only an axial deformation, but also a transversal deformation. A pencil eraser, for example, placed in and loaded vertically in compression, will be shortened in the axial direction, but the transverse dimension will increase. This phenomenon of the production of a transverse strain $\varepsilon t = -\nu\varepsilon$ is known as Poisson effect[25].

Taking into account this effect, the relationship between force and deformation is no longer linear, in fact, as the load is applied, the section area will increase approximately with the law:

$$A = A_0(1 + 2\varepsilon_t) = A_0(1 - 2\nu\varepsilon)$$

and:

$$\varepsilon = \frac{F}{EA} \approx \frac{F}{EA_0(1 - 2\nu\varepsilon)} \qquad F \approx \varepsilon \cdot EA_0(1 - 2\nu\varepsilon) = EA_0\varepsilon - 2EA_0\nu\varepsilon^2$$

therefore the real working characteristic is not linear.

---

[24] The axial deformation is the ratio between the variation of the axial length ΔL and the initial length L and is denoted by the Greek letter ε and described by the formula ε = ΔL/L.

[25] The transversal deformation ε_t for Poisson effect is appreciably proportional to the axial deformation, but has the opposite sign. The coefficient of proportionality, called Poisson's ratio, is denoted by the Greek letter *v*. The transversal deformation is described by the formula $\varepsilon_t = -v\varepsilon$

The nonlinearity theoretically determined here is always hypothetical. In reality, in addition to factors like the one discussed (theoretically here determinable), there are factors that depend on not quantifiable parameters, for which the linearity error is usually determined experimentally, as shown in Fig 3.23.

3.3.2.5 Hysteresis error

Many natural phenomena, such as the behaviour of elastic materials, present trends similar to those of the backlash error, but distributed on a multiplicity of parts which do not act simultaneously. So the overall effect is affected by the random combination of individual components and it is developed in a total contribution, at the beginning small, that will gradually increases to a maximum beyond which returns to decrease to practically disappear.

Such a phenomenon describes a spindle shape, well-known for the phenomena of hysteresis of material, both elastic and magnetic hysteresis.

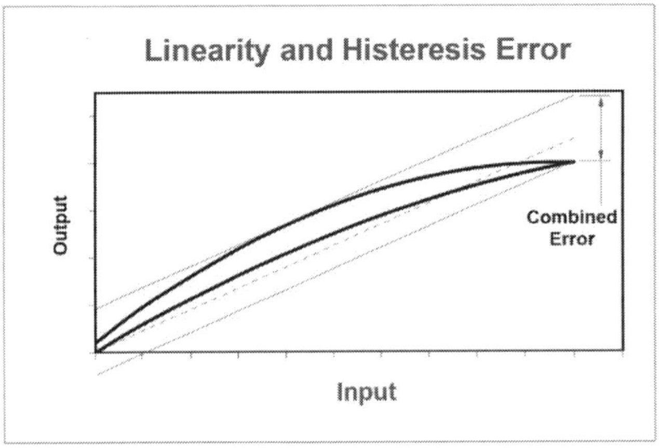

Fig. 3.23 Representation of the linearity and hysteresis errors.

Measuring instruments of any type, based on mechanical, electrical and electronic components, have in general a behaviour of this type, when there are no overriding factors like those described for the backlash error. The hysteresis effect (Fig. 3.23) is determined experimentally by results of the entire transition cycle, from the initial zero to full scale and back to the zero value. The characteristic thus obtained describes linearity and hysteresis errors. Frequently it is considered the combined error, which is not simply the sum of the two, but it depends on the concavity of the nonlinearity that determines the composition of the two parts.

3.3.2.6 Zero Error

It is often forgotten that a measurement consists in determining the difference between two conditions: the position of the first and the last points of a segment, the mass of an empty container and a full one, the deformation before applying a force and after having applied it. Many instruments, to make more easy the determination of the difference, have a tare command, which automatically resets the indication to zero. In this regard it is appropriate to make some observations. Command of zeroing is undoubtedly convenient from a practical standpoint, but makes us lose the information on the absolute level at which the instrument operates. This may lead to consequences that should be carefully evaluated, some of which are cited in the examples that follow:

- a series of successive measurements, carried out using always at the beginning of each a tare command, can hide the fact that there is a drift, that instead could be visible if all initial measurements are expressed with their original value;
- an instrument of a known calibration characteristic in the form of equation of a degree greater than the first, cannot be corrected from the nonlinearity error with such equation if the output does not use the same zero of the scale used in the calibration phase;
- recalling that any breakdown of the measuring instruments are often reported by a shift of their natural zero, it is clear how important it is that during calibration the zero condition is well defined and the value of the natural zero reported in the certificate.

If, as mentioned, on one side the reset command has the disadvantage of losing valuable information, often, on the other side, it is advantageous because it uses a zero condition better defined than the resolution of the rest of the scale. This is evident for both analogue and digital scales:

- in case of analogue scales, the initial reading may occur at any point of a division of the scale; so it must be read by eye with the interpolation which is affected by an error linked to the scale resolution. In the case of reset, instead, the index is moved to collimate with the initial trait of the scale, so that the interpolation error is avoided;
- in case of digital scales we can often observe that the zero condition is indicated by a sudden change of sign. Reaching zero from the negative side of the scale, we will observe a series of values preceded by the minus sign, until we reach zero, which is also preceded by the minus sign. Continuing to give variations in the same direction, we can observe that,, while the indication remains zero, the sign switches quickly, possibly with some fluctuations. Only after further variations also the number switches to one. The condition of switching the sign is markedly better defined than the one represented only by the presence of the zero indication.

Another contribution to the zero error, often important, is related to the measured quantity (identification of the zero condition). For example, if we have to weigh the liquid in a container, the zero condition should be determined with the container empty and dry. If we reach zero with the empty, but wet container, the zero condition is not well defined. In conclusion, from the instrumental point of view, the use of automatic zeroing can have advantages and disadvantages. Often the best way to operate is to detect periodically the natural zero of the instrument, in order to use this information it, or resetting before making the single measurements in order to improve the resolution of the zero.

3.3.2.7 Disturbance error

Sometimes the presence of the measuring instrument and the operator itself creates a disturbance on the measured quantity. The measurement variation is due to the so called *disturbance error*. An example, not quite real, but that puts in evidence this kind of error, can be given by examining the problem of measuring the temperature of a small amount of liquid.

Let's say we have few millilitres of liquid at a temperature of around 90 °C. If we want to make the temperature measurement with a normal mercury thermometer, it is evident that an energy transfer takes place from the liquid to the thermometer. Therefore, the thermometer is heated, but the liquid is cooled and, given its small amount, it cools much. We see that the result does not represent the temperature of the liquid, but an equilibrium temperature, which is lower than the original one. The difference between the balance temperature measured and the original temperature of the liquid is the disturbance error. In the described case the disturbance error is large, while usually it is small. Other examples describing how the measurement process modifies the measurand are the hardness measurements, the deformation produced by the length measurement probe on the measured object. Possible disturbance error should always be taken into account to evaluate if, and how much, it can be influent.

3.3.2.8 Error due to influence quantities

We must learn, from the consolidated experience, the information about which are the quantities, other than the main one that has to be measured, that produce effects on the measurement result. For example, in the case of load cells, instruments designed to measure the force acting in the direction of their axis, the presence of transversal forces and moments produces sometimes considerable variations in the measurement result. In the case of analytical balances, electrostatic forces generated between the weighted objects and the closing surfaces of the measurement zone produces sometimes significant variations.

The analysis of the influencing variables can be made, therefore, only knowing very well the operation principle of the used instrument.

3.3.2.9 Error due to variation in the time of the measured quantity

As has already been said, the action of measurement by an instrument takes place with an energy transfer between the measured quantity and the instrument itself. This transfer may not, of course, be instantaneous. Many measurement procedures, in general all those suitable to obtain measurements with low uncertainty, provide acquisition times sufficiently long to operate in fairly static conditions. In these, the parameters of the entire measuring complex (value of the measured quantity, conditions of the measuring instrument, effect of the presence of the operator, of environmental conditions and of other influence quantities) are constants.

In the practice of industrial measurements, this is not generally possible, because on the one hand the time taken should be minimized, on the other hand the need to operate in an environment that has constraints of production does not allow stopping everything for a long time. Moreover, sometimes, the purpose of the measurement can be just the determination of the evolution over time of the measured quantity value[26].

In all these cases, it must be remembered that the output of the measuring instrument, let's call it here y (typical notation of mathematics as dependent variable) is not systematically proportional to the input value, let's call it x (as a traditional symbol of an independent variable), when the input is a function of the time t.

Let us examine, for example, the behaviour of a common mercury thermometer when the temperature T to be measured is variable over time. If we explore the mechanism by which the measurement takes place, simplifying it, we must consider that the output is proportional to the temperature T_m of mercury and that the passage of energy dE in the time interval dt is governed by the laws of heat transmission. This it is also shown in simplified form on the basis of a thermic transmission parameter k, which takes into account the transmission coefficient and the area through which the heat is transmitted from the environment at temperature T to the mercury at temperature T_m:

$$\mathrm{d}E = k\left(T - T_m\right)\cdot \mathrm{d}t$$

the energy dE, received from mercury produces, in turn, an increase of the temperature T_m dependent on its thermic capacity C_m:

$$C_m \cdot \mathrm{d}T_m = \mathrm{d}E$$

This physical model can be used to obtain, always assuming drastic simplifications (that lead to consider constant the values of the transmission parameter k and the thermic capacity of mercury C_m), the differential equation that describes the evolution of the output as a function of the input:

$$\frac{C_m}{k}\cdot\frac{\mathrm{d}T_m}{\mathrm{d}t} + T_m = T$$

this one, in the event of the constant of k and C_m, is a differential equation of the first order. We can write the equation with the usual notations of mathematics, taking the output T_m as a dependent variable y and the input T as the independent variable x calling τ the ratio C_m/k (considered constant):

$$\tau\cdot\frac{\mathrm{d}y}{\mathrm{d}t} + y = x$$

---

[26] Doebelin E. O., “Measurement Systems”, Mc Graw – Hill, 1990, Cap. 3.3 and Beckwith T. G., Marangoni R. D., Lienhard J. H., “Mechanical Measurement”, Addison Wesley, Fifth edition, Cap. 4

of which the solutions are known. For example, the usual operation of dipping the thermometer in a fluid, to measure its temperature, produces a sudden change $\Delta T = T\text{-}T_0$ of T that passes from room temperature T_0 to that of the fluid T. Sudden change can be represented by a step function. Let's examine the case of the origin of the times taken in the instant of immersion:

$$\begin{cases} t<0 & x=0 \\ t\geq 0 & x=\Delta T \end{cases}$$

We take as dependent variable y, the variation of the temperature of the mercury compared to its initial temperature T_0:

$$y = T_m - T_0$$

in that case the solution is:

$$y = x\left(1-e^{-t/\tau}\right) = \Delta T\left(1-e^{-t/\tau}\right) \qquad T_m - T_0 = \left(T-T_0\right)\left(1-e^{-t/\tau}\right)$$

and the difference:

$$T - T_m = \left(T-T_0\right)e^{-t/\tau} = \Delta T \cdot e^{-t/\tau}$$

is the error due to the time variation of the measured quantity. When we take temperature measurements in the described way, we have to wait an adequate time before doing the reading. The trend of the output is asymptotic compared to the value to be measured, as shown in Figure 3.24.

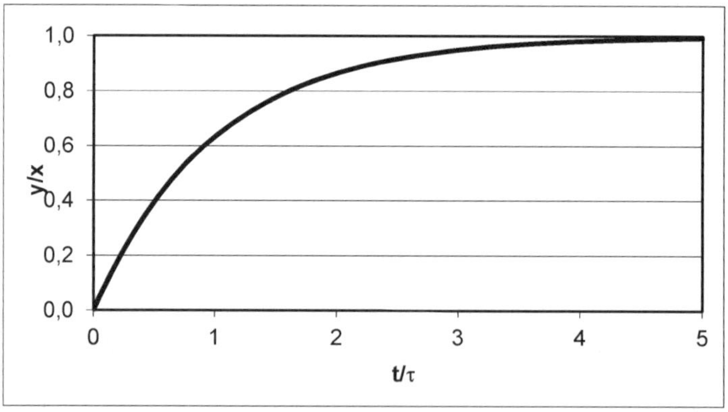

Fig. 3.24 Trend of the ratio output/measured quantity by an instrument of the first order when the measured quantity undergoes a step variation.

In base of the common experience, however, it is easy to realize that not all the instruments behave as described above. In fact, for example, balances do not reach asymptotically the measurement value, but they reach it oscillating around it.

Let's try to examine the factors that are involved in their functioning. It is advantageous to think in terms of force, rather than in terms of moment[27], since the form of the equation does not change and the symbols are simpler.

We will consider a force transducers (i.e. a spring force transducers) subjected to an equilibrium force F_e variable over time, a restore force F_r that tends to bring the load receptor back to the balance position, a viscous force F_v that slows down the movement and, finally, an inertia force F_i. If the load receptor gets moved to the position s from its equilibrium position s_e, a restore force F_r, proportional to the movement imposed, is created and it can be calculated using the elastic constant k defined by:

$$k = \frac{\Delta F}{\Delta s}$$

Therefore the elastic force in the equilibrium position and the restore force are given by:

$$F_e = k \cdot s_e$$

$$F_r = k \cdot (s - s_e) = ks - F_e$$

The viscous force F_v will be proportionate to the speed of displacement:

$$F_v = h \cdot \frac{\mathrm{d}s}{\mathrm{d}t}$$

and, because the moving parts have their mass m, it is created an inertia force F_i by the accelerations of the various parts, proportional to the second derivative of the displacement versus time:

$$F_i = m \cdot \frac{\mathrm{d}^2 s}{\mathrm{d}t^2}$$

For the equilibrium, the sum of these three forces has to be null, therefore:

$$F_i + F_v + F_r = 0$$

$$m \cdot \frac{\mathrm{d}^2 s}{\mathrm{d}t^2} + h \cdot \frac{\mathrm{d}s}{\mathrm{d}t} + k \cdot (s - s_e) = 0$$

The output $\boldsymbol{u}$ can be considered proportional to the displacement, with a proportionality transducer constant $\boldsymbol{k_1}$ that define the relation between movement and output:

$$s = k_1 \cdot u$$

and, therefore, we obtain the relation between the signal variation $\boldsymbol{u}$ and the

[27] For balances with equal arms, in which the movement is a rotation of the beam, we should write an equation of moments.

equilibrium signal u_e, from which is possible to determine the error due to the variation of the time of F_e:

$$m \cdot \frac{d^2 k_1 u}{dt^2} + h \cdot \frac{dk_1 u}{dt} + k \cdot (k_1 u - k_1 u_e) = 0$$

$$m \cdot k_1 \frac{d^2 u}{dt^2} + h \cdot k_1 \frac{du}{dt} + k \cdot k_1 (u -_1 u_e) = 0$$

$$m \cdot \frac{d^2 u}{dt^2} + h \cdot \frac{du}{dt} + k \cdot u = k \cdot u_e = F_e$$

The previous formula, if written with the mathematics notations, using the output, u, as dependent variable, y, and the input, F_e, as independent variable, x, is a differential equation of the second order like:

$$m \cdot \frac{d^2 y}{dt^2} + h \cdot \frac{dy}{dt} + k \cdot y = x$$

Let's examine the first solution in case x is a sinusoidal function:

$$x = a \sin(\omega t)$$

we define the natural angular frequency of the system: $\omega_0 = \sqrt{\frac{k}{m}}$

the critical damping: $h_c = 2m\omega_0$

and the equilibrium output: $y_e = \frac{x}{k}$

The steady solution, represented in Fig 3.25, is:

$$\frac{y}{y_e} = \frac{1}{\sqrt{\left(1 - \frac{\omega^2}{\omega_0^2}\right)^2 + 4\left(\frac{h\omega}{h_c \omega_0}\right)^2}} \sin(\omega t + \phi)$$

where the phase shift ϕ, represented in Fig 3.26, is given by:

$$\tan \phi = -2 \frac{h \omega \omega_0}{h_c (\omega_0^2 - \omega^2)}$$

The diagram of the relationship between the amplitudes of the instantaneous output and the output at equilibrium shows well the magnitude of the error due to the variation in time of the measured quantity. In particular, it can be observed that when the oscillating frequency of the measured quantity approaches the frequency of the system, the error grows strongly if the damping is not high enough.

In addition to the error in amplitude, we must take into account the phase shift of the output compared to the input; in other words, in addition to the amplitude variation, there is also a movement over time which can represent also a measurement error if we are interested to know the time mode of the analysed phenomenon.

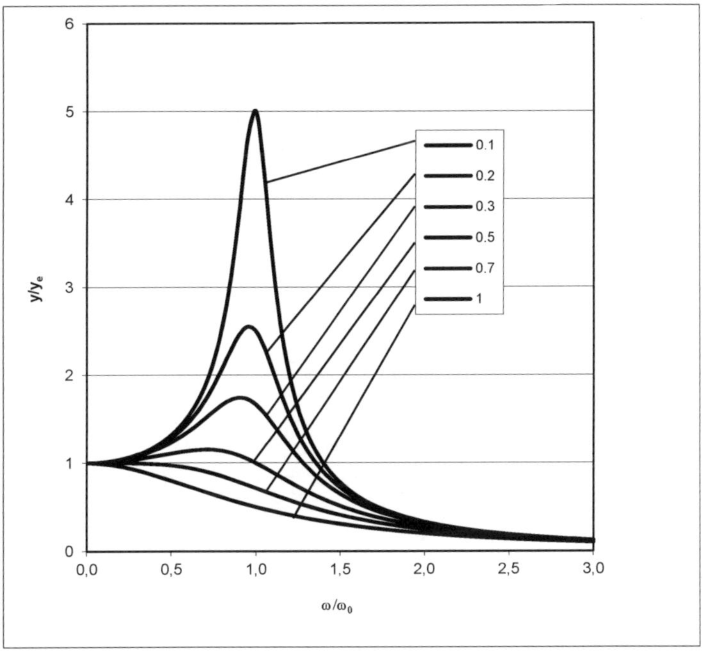

Fig. 3.25 Shape of the relationship between the output and the output of balance, for values of the ratio h/h_c between damping and critical damping varying from 0.1 to 1.

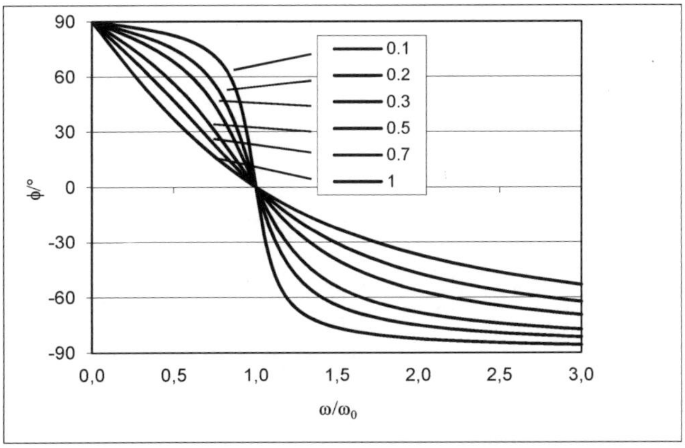

Fig. 3.26 diagram of the phase shift ϕ in degrees, for values of the ratio h/h_c between damping and critical damping variable from 0.1 to 1.

The answer to the sinusoidal function is important since, in practical cases, the possible trends in time of the values of the measured quantity can be represented by a Fourier series which divides the input into its harmonic components.

It is obvious that, if for the different harmonics the relationship between input and output is different, the shape of the output will be distorted compared to the input. The distortion is acceptable if the higher harmonic, still present in the input, is contained in the frequencies field for which the ratio y/y_e is sufficiently close to 1.

Returning to our force transducers, the immediate application of a load is, however, a step function x:

$$\begin{cases} t < 0 & x = 0 \\ t \geq 0 & x = F_e \end{cases}$$

Even in this case it is known the solution of the differential equation. In case the damping is smaller than the critical damping, (i.e. $\xi = \boldsymbol{h/h_c} < 1$) we have:

$$\frac{y}{y_e} = 1 - \frac{e^{-\xi\omega_0 t}}{2}\left[\cos\left(\omega_0 t\sqrt{1-\xi^2}\right) + \frac{\xi}{\sqrt{1-\xi^2}}\sin\left(\omega_0 t\sqrt{1-\xi^2}\right)\right]$$

Instead, in the case the damping is bigger than the critical damping we have:

$$\frac{y}{y_e} = 1 - \frac{e^{-\xi\omega_0 t}}{2}\left[\left(\frac{\xi}{\sqrt{\xi^2-1}} + 1\right)e^{\omega_0 t\sqrt{\xi^2-1}} - \left(\frac{\xi}{\sqrt{\xi^2-1}} - 1\right)e^{\omega_0 t\sqrt{\xi^2-1}}\right]$$

Let's see what happens to our force transducers (Fig. 3.27).

Let's assume that the equivalent of the moving mass is 3 kg, the stroke of the plate, when a mass of 1 kg is applied, is approximately 2 mm, then the stiffness $\boldsymbol{k}$ is approximated to 5 N/mm (we do here only an approximate calculation to see an indication of the trend).

From the above data we can calculate the natural rotating frequency:

$$\omega_0 = \sqrt{\frac{k}{m}} = 40{,}8\,\mathrm{s}^{-1}$$

and the critical damping:

$$h_c = 2m\omega_0 = 245\,\mathrm{kg/s}$$

and then calculate the trend of the signal for different damping values, defined by the ratio ξ. The results are represented in Fig 3.27.

It may be interesting to evaluate the time required for the instrument to provide the output with an error lower than an acceptable value. An instrument that quickly reaches such condition is said to be *ready*.

The readiness is strongly influenced by the damping. If the damping is low ($\xi = 0.2$), the instrument quickly reaches its position of balance, but surpasses it abundantly and continues to oscillate. Therefore it requires a high amount of time to keep continuously in the desired range with the condition of balance.

If the damping is high, the instrument reaches the balance condition in an asymptotic way, but it takes a long time to enter in the desired range.

Thus, especially when the acceptable error is low, the most favourable condition for quickly reaching the balance condition and remain in a small range of it, is a damping, with ξ near to 0.8.

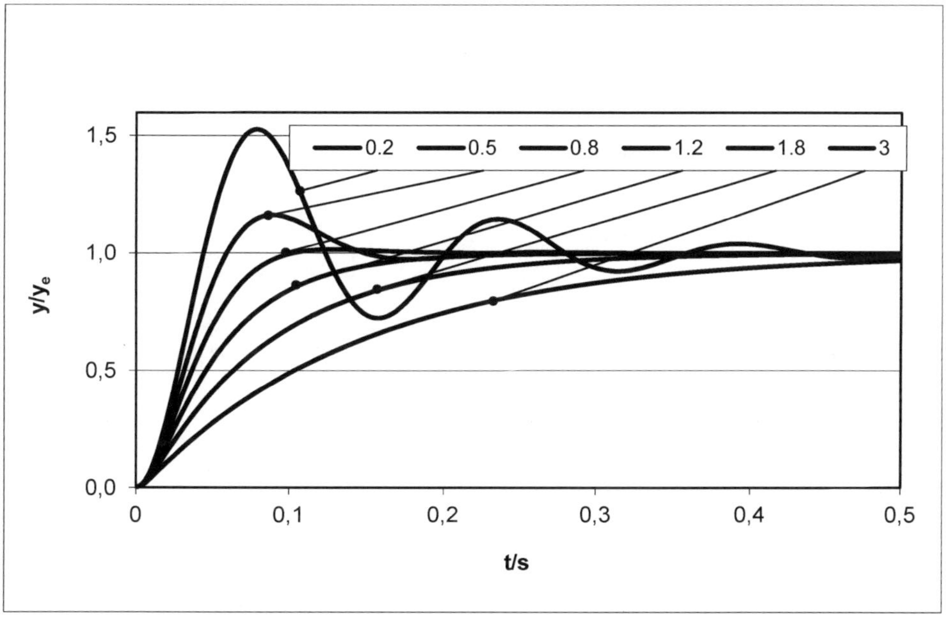

Fig. 3.27 Reaching of the balance position of an instrument of the 2nd order to the change of the damping here defined by the ratio ξ with the critical damping.

4 BEYOND DETERMINISM: NOTES ON PROBABILITY

4.1 DETERMINISM AND RANDOMNESS

The mathematical experience in school education starts, usually, by a strictly deterministic organization, that is based on exact relationships between data and results. The methods studied, from the multiplication tables up to the development of differential and integral calculus, do not allow alternatives, if not equivalent.

This gives the reason why mathematics is considered an "exact science". Coming to practice, the exactitude does not exist: we cannot get an exact result starting from doubtful data! This limit might make mathematics unusable because the results of exact calculus do not correspond to reality, but that would deprive us of a powerful means for the management of many technical issues.

In practice the trick consists in using the exact calculus over doubtful data by assigning a conventional variability to the result. For example, in the design of building structures, as a house, a bridge or a dam, we carry out deterministic calculations with mathematical rules, but, at the same time, we apply the so-called "safety coefficient", i.e. the data (for example, the expected loads) or the results (e.g., size of columns) are modified by a factor that accounts for the effects of ignorance of the exact value of the data used (we do not known what will be the actual loads, nor what is the actual breaking strength of materials) or irregularity of implementation (for example, the proportions of the components of the concrete and size of beams and columns may not be accurate).

This approach is common in many fields: for example, in the mechanical sector it is adopted the system of dimensional tolerances, as products cannot have all the exact nominal dimension, and in legal metrology it is defined conventionally for each measuring instrument the value of an accepted *Maximum Permissible Error* (*MPE*). These factors are established, generally, on a conventional base, using the feedback of experience. This, however, applies only to situations well framed in usual schemes and requires large margins to ensure coverage in all those cases, sometimes sporadic, that may arise.

In metrological field and in many other fields, these two conditions are not always present.

Production requires a specific process for each different product, often not referable to a standardized procedure. Standardization bodies (ISO, ASME, etc.) make a lot of efforts, but they can deal only when consolidated evidence from long experience has made the procedure conventional[1]. Many situations in

[1] For example, the tensile test of materials is normalized as forms of specimens and test conditions for all those situations, even complex, where you can create a contact sheet of definite shape. It cannot be normalized, if not with great difficulty, for tests on complete facilities, which can have different shapes and are not easily traceable to a conventional pattern.

all areas of application, from technology to economy, require a treatment of non-deterministic specifications, so with the help of mathematics, there were developed techniques of combinatorics, probability and statistics that help we deal with problems more commensurate with the *random* nature.

This text does not want to give a detailed treatment of probability calculus, but only provide the basics needed to understand the next part of statistics and applications to design, production and verification industrial practice. The reader who wants to learn more can refer to specific texts. There, in fact, many of the arguments, presented here in a simplified form functional to the purposes of this text, are treated in rigorous mathematical mode with a much wider study of probabilistic and statistical part.

4.2 THE USE OF PROBABILITIES

To overcome the limits of determinism is very useful to address problems using probabilistic concepts. As mentioned before, in real practical applications the result cannot be determined exactly since data are not exact, so that the entry variables in our formula are not exact, but randomly changing values; these can be called *random variables*. In this case we cannot get a defined value for the result, but we can only estimate the probability that it falls within a certain range of values.

We must, therefore, understand what is meant by the term "probability". We will examine two forms of the concept of probability, taking as example what happens rolling a six-sided dice:

- If we rely on an assumption of homogeneity and symmetry between sides, it is evident that it cannot be any difference among all the sides, therefore the probability of getting anyone of the sides is equally subdivided between the sides and, for a six-sided dice, it is 1/6. Instead, for a coin we have only two sides, the probability of getting each side is 1/2.
  In general we can give the so called classical definition of probability as the ratio between the number of equivalent selected events (if we bet on the dice side with 3, we have only one winning event; if we can bet at the same time on the side with 3 and on the side with 4, we have two winning events) to the total number of possible events (six in case of six-sided dice, two in case of a coin).
  Rolling a six-sided dice, possible outcomes are 1, 2, 3, 4, 5, and 6. The set of possible outcomes is called "sample space", frequently indicated as S = {1,2,3,4,5,6}.
  Cardinality[2] of the sample space #S, is the number of equivalent events that can outcome (six in this case). If you bet on 2 and 4 the set of expected events can be indicated for example by E = {2, 4}, and cardinality of E is #E=2. Your probability to win is #E/#S, that is, 2/6.

[2] The sign #put before the symbol of a set indicates its cardinality, i.e. the number of its elements.

- In many cases, a theoretical assessment as the previous one is very difficult, if not impossible. The mere fact that the dice is unbalanced or with unequal sides makes impossible the previous reasoning: resort to symmetry is not possible. We can try, then, to evaluate the probability "a posteriori", rolling many times our dice considering how many times we get our expected event in reference to the total number of trials. This way of defining the probability is frequently called frequentist concept of probability.

Notice the deep differences between these two concepts.

The first, entirely hypothetical, based on acceptance of conditions of equality and homogeneity between situations that, in practice, can be different, and not homogeneous. Nevertheless it is very efficient in generating theoretical settings that allow us to reach, without heavy experimental work, conceptual calculation schemes.

The second addresses, on an experimental basis, problems that would be very difficult to deal in theory; just think what complications of assumptions and calculations could be introduced in simple probability assessment linked to the individual faces of a *real* pseudo-cubic, only nearly balanced dice.

4.2.1 The concept of frequentist probability

It is important to underline concrete implications, not just hypothetical, of probability assessment in real-world situations, because in this way we can highlight two things:

- the advantage of frequentist concept consists in involving real conditions. In practice the assessment of probabilities depend often by factors not easily evaluated theoretically (e.g. for the dice, the position of the mass centre or the local geometry of fillets and concavity or convexity of sides, that are taken into account by rolling our dice);

- the drawback of frequentist concept is that it requires a significant amount of experimental work, with all the associated problems of cost, time and, not least, of dependence on influencing factors introduced by the experimental procedure.

The advantage of the method of evaluating the frequentist probability is that it does not need any symmetry among the different possible events, therefore it can be applied to complicated phenomena. The probability of a complex event, then, is usually estimated by making a series of tests and calculating the relative frequency (number of selected results divided by the total number of tests) as the corresponding probability estimation.

Note that a good estimation requires a high number of tests. For example, let's look at the distribution of results in throwing a dice with 50, 200, 500 and 1000 trials (fig. 4.1). We see how 50 tests[3] are generally inadequate to accurately assess the probability (the data collected would indicate a significant difference in probability between the different sides), while the data calculated with 200 or more tests represent adequately the probability distributions, confirming the theoretical value of 1/6 for each side.

If we accept an experimental activity quite extended, we can treat also greatly complicated cases.

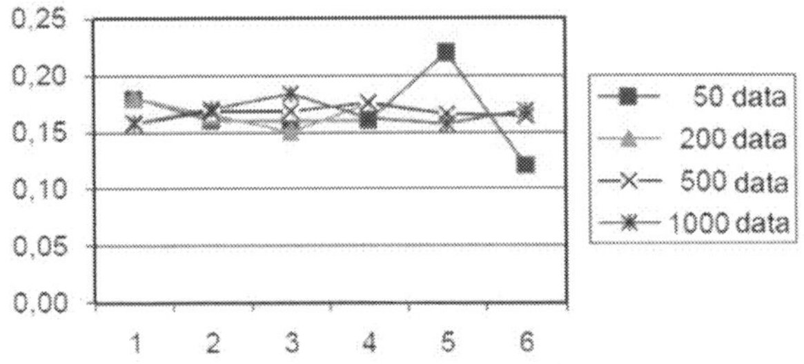

Fig. 4.1 Results obtained with the launch of a six-sided dice.

The practical difficulties linked to a correct preparation of the experiment and its execution, that shall be replicated many times, indicate that the actual usefulness of frequency method is limited to the determination of the probability of the events for which it is not possible to give an estimate based on homogeneity and symmetry considerations.

4.2.2 The classical concept of probability

Many examples related to the classical concept of probabilities lead us to gambling. We might ask ourselves what is the probability to retrieve the two of Spades from a deck of 52 cards, or that, throwing two dice, the sum is 8 etc.

To assess their probability based on the classical concept of probability, we must try to estimate the number of selected cases and the total number of cases that may arise.

In the example of the deck of cards, the extraction can lead to 52 different results, as many as there are cards in the deck, but among these only 1, the two of Spades, is that considered selected, so we can say that we have 1 chance on 52.

To make this calculation we must, therefore, determine the set of all possible outcomes. It is often denoted by the letter **S** (*Sample Space*) and called "sample space". In the case of deck of 52 cards, sample space is represented by the 52 card values.

[3] Note that, in many cases, 50 tests are an unattainable number both in scientific research, as well as in industrial practice, and this shows that it is not only useful, but indispensable, in certain situations, working on hypothetical basis with theoretical estimates.

Do not misunderstand that the sample space is related to the number of objects (the 52 cards in this case); it relies to the number of different possible events. If we consider that each card is different from all the others, that we lose if we bet on the 2 of spades, and the 2 of hearts arrives , we really face 52 possible events (ace of spades (♠), ace of hearts (♥), ace of diamonds (♦), ace of clubs (♣), two of spades (♠), two of hearts (♥), etc.). Therefore, the cardinality #S of the sample space is 52. But, if we bet to getting a 2, disregarding the suit, the possible events are only 13 (ace, 2, 3, …, J, Q, K) so #S=13. In the same way, if we bet on an ace, we have our expected event set E with a cardinality #E=1 and our probability to win is 1/13. If we consider only the colour, as there are two colours in our deck, our sample space cardinality is #S=2. We can see that the sample space is not relied to the number of objects (always 52 in the deck), but to the number of different possible conditions. Note that the question "what is the probability of getting a red card?" can be solved in two ways: 1) the sample space cardinality for colours is 2, the winning event cardinality is 1, so the winning probability is 1/2; 2) considering all cards with a cardinality of 52, since the winning event is to take one of the 26 red cards, the cardinality of our winning event is 26, therefore the winning probability is 26/52=1/2.

In the case of two dice, if we bet on the sum of the two outcomes, we can already arrange the sample space as in fig. 4.2, according to the value of the sum. We can observe that the number of possible outcomes is 36 and that the number of selected results, e.g. those who give sum 8, is 5. The probability of getting a sum of 8 is, therefore, of 5/36.

						6-1					
					5-1	5-2	6-2				
				4-1	4-2	4-3	5-3	6-3			
			3-1	3-2	3-3	3-4	4-4	5-4	6-4		
		2-1	2-2	2-3	2-4	2-5	3-5	4-5	5-5	6-5	
Outcomes	1-1	1-2	1-3	1-4	1-5	1-6	2-6	3-6	4-6	5-6	6-6
Sum	2	3	4	5	6	7	8	9	10	11	12

Fig. 4.2 Representation of the sample space of the outcomes of two 6-sided dice.

Establish a definition of selected events **E** (in the first case the 2 of Spades, in the second the sum equal to 8) and consider the corresponding number **e** = **#E** of selected events and the total number **s** = **#S** of events in the sample space. The probability $\mathcal{P}$(**E**) of getting one selected event is given by the ratio **e/s** = **#E/#S**.

In practice we can see that the problem consists in correctly defining the sets **E** and **S**. A good help may come from combinatorics.

4.2.2.1 Tips on Combinatorics

The table represented in fig. 4.2 can be written just because the Sample Space is small, but if we try to represent the Sample Space of rolling of three dice it starts to be difficult to keep it in one page; for four dice it is nearly impossible. Note that the real cases involving production processes can be even more complex.

If we examine the case of throwing two dice, we can say that the sample space consists of all possible pairs of numbers from 1 to 6.

Let's break this problem into easier parts, beginning to wonder in how many ways we can sort a group of n objects.

Let us start by two objects, which we will call **a** and **b**. In this case, the answer is simple, because we can put them in the succession **a**, **b** or **b**, **a**, so we only have two different situations which, as we can see, we got permuting the order of objects and, hence, are called "*permutations*".

We now come to three objects, **a**, **b** and **c**.

Let us arrange the permutations in this way: first we take **a** fixed and we commute **b** and **c**, then we keep **b** fixed and we commute **a** and **c**; finally, we take **c** fixed and we commute **a** and **b**:

1	**a, b, c**	3	**b, a, c**	5	**c, a, b**
2	**a, c, b**	4	**b, c, a**	6	**c, b, a**

We can observe that the number of permutations of three objects (6) is obtained considering each object as many times as the number of permutations of two objects. Therefore the number of permutations of three objects can be calculated multiplying the number of the objects n=3 for the number of permutations of n-1=2 objects. If we try with four objects we can note that this is a general condition, therefore this allows us to establish the general rule:

The permutations of n objects are given by permutations of n -1 objects multiplied by n

If we take into account the evident fact that the permutation of one object is 1, we obtain a general formula:

permutation of one object = 1
permutation of two objects = $1 \cdot 2$
permutation of three objects = $1 \cdot 2 \cdot 3$
......
permutation of n objects = $1 \cdot 2 \cdot 3 \ldots n = \prod_{1}^{n} i = n!$

The number of permutations of n objects is given by the product of the integers from 1 to n, which is called *factorial* and symbolized with an exclamation point after the number.

Take a look at the permutations of four objects **a**, **b**, **c**, **d**:

1	**a, b, c, d**	7	**b, a, c, d**	13	**c, a, b, d**	19	**d, a, b, c**
2	**a, b, d, c**	8	**b, a, d, c**	14	**c, a, d, b**	20	**d, a, c, b**
3	**a, c, b, d**	9	**b, c, a, d**	15	**c, b, a, d**	21	**d, b, a, c**
4	**a, c, d, b**	10	**b, c, d, a**	16	**c, b, d, a**	22	**d, b, c, a**
5	**a, d, b, c**	11	**b, d, a, c**	17	**c, d, a, b**	23	**d, c, a, b**
6	**a, d, c, b**	12	**b, d, c, a**	18	**c, d, b, a**	24	**d, c, b, a**

Tab. 4.1 Outline of permutations of four objects

If we want to see how many ways two objects (k = 2) can be extracted from a group of four objects (n = 4), we can consider the first two items in each of the permutations. We see that, for example, permutations 1, 2, 7 and 8 contain all the objects **a** and **b** in the first two positions, albeit in different order. But frequently we are interested in the objects contained in the group, not in their order[4]; these four groups are equivalent in reference to the winning conditions, therefore shall be considered as a single one in the sample space. This condition evaluated for **a** and **b** is completely general, and it can be seen, for instance, for **a** and **c** in permutations 3, 4, 13 and 14.

The number of times such equivalence occurs is given:

- for the equivalence between the permutations 1 and 7 and 2 and 8, the permutations of k = 2 objects **a** and **b** (only the exchange from **a**, **b** to **b**, **a**);
- for the equivalence between the permutations 1 and 2 and 7 and 8, the permutations of n - k = 2 remaining objects (in this case, **c** and **d**).

The number of permutations, equivalent for containing always the same objects in the first k places, will then be calculated as the product of permutations of k objects times the permutations of n - k objects, i.e. for the example with n = 2 and k = 2:

$$k!\cdot(n-k)! = 2!\cdot 2! = 4.$$

Considering this, it is easy to calculate the number of groups really different for the first k objects involved, not only for their order, called "*combinations*".

The formula that calculates combinations, using conventional symbols, is shown here:

$$\mathrm{Comb}(n,k) = \frac{n!}{k!(n-k)!} = \binom{n}{k} \qquad (4.1)$$

---

[4] If you bet to draw the ace of hearts from a deck of cards, the important thing is that the first card you get is the ace of hearts. If you bet to draw the aces of hearts and clubs, it is important that the first two cards you get are these two aces, no matter if first hearts or first clubs.

in the case of the example, n = 4 and k = 2, so the number of combinations is 6:

$$\mathrm{Comb}(4,\ 2) = \binom{4}{2} = \frac{4!}{2!(4-2)!} = 6$$

as we can easily check, putting our objects in the alphabetical order, for avoiding to have groups with different order:

1	a, b	3	a, d	5	b, d
2	a, c	4	b, c	6	c, d

Example. Calculate and check the combinations of four objects three at a time. In this case we have n = 4 and k = 3, so the formula is easy:

$$\frac{n!}{k!(n-k)!} = \frac{4!}{3!(4-3)!} = \frac{24}{6 \cdot 1} = 4$$

In fact, using the permutations of four objects (table 4.1), grouped together in order to have the first three equivalent (from 1 to 6 always objects **a**, **b**, **c**, from 7 to 12 objects **a**, **b**, **d**, from 13 to 18 **a**, **c**, **d**, and eventually from 19 to 24 items **b**, **c**, **d**), we can see that the possible combinations are:

a, b, c   a, b, d   a, c, d   b, c, d

With the use of combinatorics, checked here on simple examples, we can rapidly calculate also schemes that are not easily representable on a paper page. For example, *Lotto* classical game draws 5 objects from 90 objects. The Sample Space cardinality **#S**, is the number of different groups of 5 objects that can be drawn from 90 objects, that is combinations of 90 objects at 5 at a time.

$$\frac{n!}{k!(n-k)!} = \frac{90!}{5!(90-5)!} = 43\ 949\ 268$$

It is recommend that we do not try to verify this result writing, as before, all possible combinations, because it would require a fairly long job!

The evaluation of the Event Space is usually not simple. Try for example to evaluate to the probability of a "*Terno*", that is the bet on three objects and the win occurs if they are among the five objects drawn.

Again the way of solution is to work on an example with small numbers: we can consider a total number of objects n = 6, a group drawn of k = 4 and that we bet on m = 2 objects.

The Sample Space is simple: its cardinality is calculated by the combinations of 6 objects at 4 at a time:

$$\frac{n!}{k!(n-k)!} = \frac{6!}{4!(6-4)!} = 15$$

As can be seen here in alphabetical order:

abcd	abce	abcf	abde	abdf
abef	acde	acdf	acef	adef
bcde	bcdf	bcef	bdef	cdef

if we bet on the objects **a** and **b**, the first six groups let we win (check that there are six winning groups for any couple of objects).

Why six? How can we calculate it?

Look at the other objects after **a** and **b**, that is **c**, **d**, **e** and **f**. We can see that they are get organized as their combinations:

cd ce cf de df ef

Therefore calculating the combinations of $n - m$ objects at $k - m$ at a time we evaluate the cardinality of the event space:

$$\binom{n-m}{k-m} = \frac{(n-m)!}{(k-m)! \cdot [(n-m)-(k-m)]!} = \frac{4!}{2! \cdot 2!} = 6$$

This allows us to evaluate the Event Space for a “Terno” in Lotto, using the relevant values $n = 90$, $k = 5$ and we bet on $m = 3$ objects:

$$\binom{n-m}{k-m} = \frac{(n-m)!}{(k-m)! \cdot [(n-m)-(k-m)]!} = \frac{87!}{2! \cdot 85!} = 3741$$

so that the probability of winning a “Terno” at Lotto is 3741/43 949 268 $\approx 8.5 \cdot 10^{-5}$.

We certainly noticed that the evaluation of the Event Space is not simple and it requires to work carefully on a small example looking for organizations of groups that can be evaluated by combinatorics.

4.2.3 Compound Probability

The “simple” probability situation described above allows to treat only cases for which we can calculate the cardinalities of Sample Space and Event Space. When more than one Event is involved, the scheme is more complicated; for instance:

- rolling twice a six-sided dice, what is the probability of getting twice the five?
- If we drew from the deck of 52 cards two aces, what is the probability, drawing out three more cards, to get the other two aces?
- If we drew from the deck of 52 cards four clubs suit cards in scale from 5 to 8, what is the probability, drawing a fifth card, to get the clubs suit or a 9?

We certainly can solve these problems with combinatorics, but it is not simple. A more practical way is to adopt the so called “tree diagram”, because it assumes the shape of a tree with branches which are subdivided at each successive step.

The preparation method is simple: we subdivide our problem in all the successive steps. In the first case we have two successive rolling of a dice (the rolling of two dice is equivalent). The first roll can leave six different results, of which only one, the five, is selected. There are two alternatives:

- we get the 5, and this happens once out of six;
- we get an outcome different from 5, and this happens five times out of six.

We start drawing a fork that can be represented by two branches: one, corresponding to the 5 with probability 1/6, the other, corresponding to an outcome different from 5, with probability 5/6. It can be seen in the diagram of fig. 4.3.

Since the selected condition requires to get 5 twice in two rolls, we take into account only the branch corresponding to the success, because if we fail the first roll no result of the second one is suitable. From that point we have, again, the same alternative as before: in the second roll there is a possibility to get 5 with probability 1/6, or to get an outcome different from 5 with probability 5/6. We can build the other two branches of the tree diagram as shown in fig. 4.4.

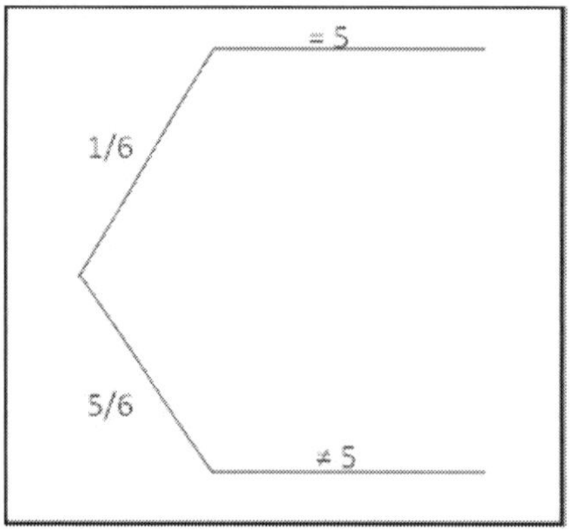

Fig. 4.3
Tree diagram to get a 5 rolling a 6-sided dice.

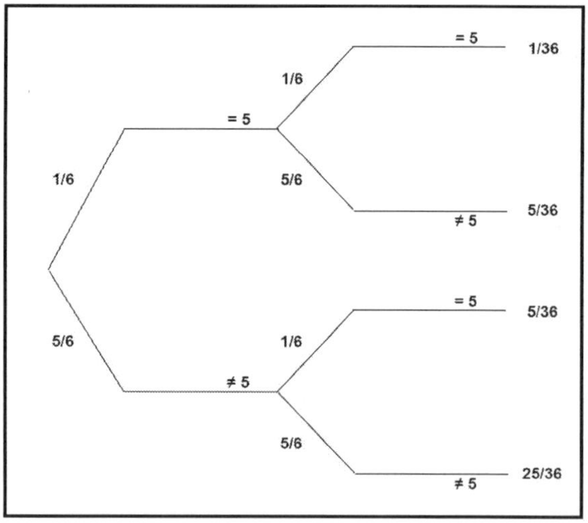

Fig. 4.4
Tree diagram to get two 5 by rolling two times a 6-sided dice.

We calculate the probability to get twice 5 as defined in classical probability. Sample Space **S** consists of all pairs that can be composed with the six outcomes of the first and the second rolls as shown in table 4.2. Clearly the number of elements in the sample (in the rectangular box) is 36. In table 4.2 is also highlighted the double-outcome of 5, representing the selected event that occurs only once. The corresponding probability is 1/36.

The same calculation can be done much more easily with the tree diagram, following the path that corresponds to the event. In this case the path consists of two representative branches of =5, each characterized by 1/6 probability. The result is obtained by multiplying the probabilities of each of the branches paths, then:

$$(1/6)\cdot(1/6) = 1/36$$

Tab. 4.2 Sample Space for throwing two dice

		Second dice					
		1	2	3	4	5	6
First dice	1	1-1	1-2	1-3	1-4	1-5	1-6
	2	2-1	2-2	2-3	2-4	2-5	2-6
	3	3-1	3-2	3-3	3-4	3-5	3-6
	4	4-1	4-2	4-3	4-4	4-5	4-6
	5	5-1	5-2	5-3	5-4	5-5	5-6
	6	6-1	6-2	6-3	6-4	6-5	6-6

So, if we wanted to calculate the probability that the outcome is 5 on the first roll and ≠ 5 on the second roll, we follow the path corresponding to these conditions having a probability of 1/6 for the first roll and 5/6 for the second one, so that the probability of this new selected event is 5/36, as we can check in table 4.2.

We can also verify with the tree diagram that if the selected event is to get only one 5 with the two rolls we can follow two branches: one getting a 5 with the first roll and ≠ 5 with the second one; the alternative is getting ≠ 5 with the first roll and 5 with the second one. This second branch has the same probability 5/36 as that before, but for us the two branches are equally suitable, therefore the total probability is their sum, 10/36, as we can check in table 4.2.

The classical definition of probability allows us to arrive at this conclusion, which can be summed up in these two indications:

- the probability corresponding to a branch of a tree diagram is given by the product of the probabilities of each step of the branch itself;
- the total probability of different branches that lead to the same selected event is given by the sum of the relevant probabilities.
- Let's solve now the second problem:
- If we drew from the deck of 52 cards two aces, what is the probability, drawing out three more cards, to get the other two Aces?

Having drawn the two Aces, they remain in the deck other two Aces and a total of 50 cards. After each draw the number of cards decreases by one (extraction without resampling).

This procedure can be represented by a tree diagram, as shown in fig. 4.5.

Still remains true that the probability for each path is given by the product of the individual probabilities, paying attention to the fact that, because each time the Sample Space changes, also the corresponding probability changes. The total probability is given by the sum of the probabilities of the three possible paths:

$$\mathcal{P} = (2/50)\cdot(1/49) + (2/50)\cdot(48/49)\cdot(1/48) + (48/50)\cdot(2/49)\cdot(1/48)$$

Apart from greater complexity related to the possible sequence of events, this problem was solved in the same way as described for the first problem.

We examine, finally the third problem:

- If we drew from the deck of 52 cards four clubs suit cards in scale from 5 to 8, what is the probability, drawing a fifth card, to get the clubs suit or a 9?

We can now see that, in this case, we can't draw a tree diagram that corresponds to the scheme of the question, i.e. with two paths, linked to two selected events: to get a clubs suit or to get a 9.

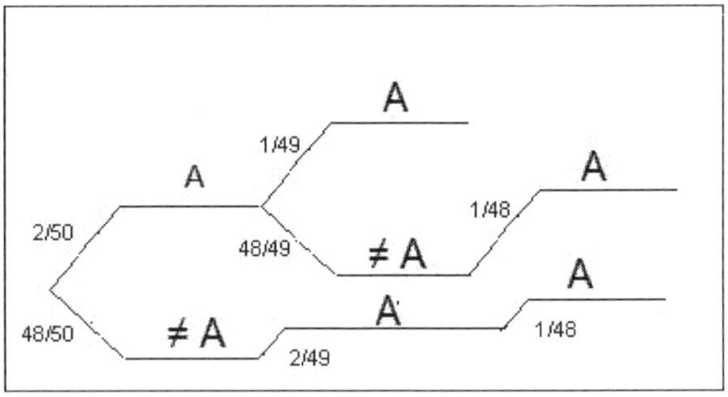

Fig.4.5 You have two cards which are two aces. Tree diagram to get the other two aces drawing three other cards.

In fact, if it was set up a branch for the "... clubs suit ..." it would also contain the 9 of clubs, then a condition belonging to the second event. Similarly, a branch set on getting a 9 would also contain a clubs suit card, the 9 of clubs. One can see that the two events given before are not mutually exclusive and this does not allow us to describe the situation with a tree diagram with just two branches.

To make the representation with the tree diagram we must break down the descriptions of events to make them incompatible. In the case exemplified, we can isolate the mixed condition 9 of clubs, then have three conditions: "... clubs suit but not 9 ...", "... 9 but not clubs suit... "and" ... 9 of clubs... " (fig.4.6).

The probability of the compound selected event of the third problem is, therefore, the sum of the probabilities of the three branches that lead to selected condition:

$$\mathcal{P} = 8/48 + 3/48 + 1/48 = 12/48$$

A visual representation which highlights clearly the condition produced by the presence of incompatible events can be made with *Venn diagrams* (fig. 4.7).

In fact, the Sample Space **S** represents the set of possible events, while Event Space **E** represents the set of selected events. The probability of a selected event will be given, as mentioned before, by the ratio of the cardinality **#E** and the cardinality **#S**.

If we require the probability of occurrence either event **E** or event **F** incompatible, the sets that represent these events are disjoint, because those events have no element in common (fig. 4.8).

The cardinality of selected events is then the sum of cardinalities of events **E** and **F**. As a result, the total probability is the sum of the two respective probabilities.

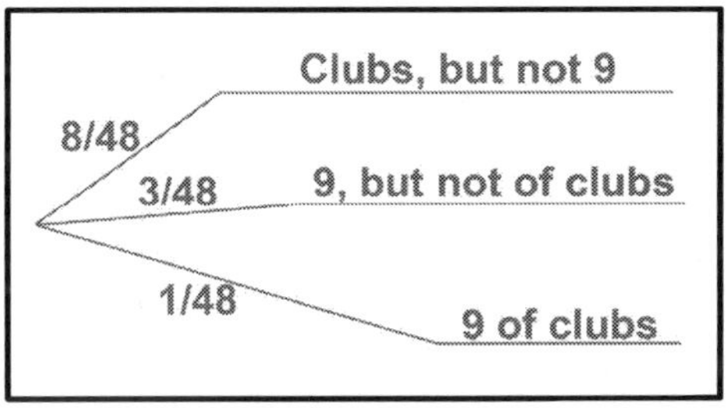

Fig. 4.6 You have 5, 6, 7 and 8 of clubs. Tree diagram to get a card of clubs or a 9.

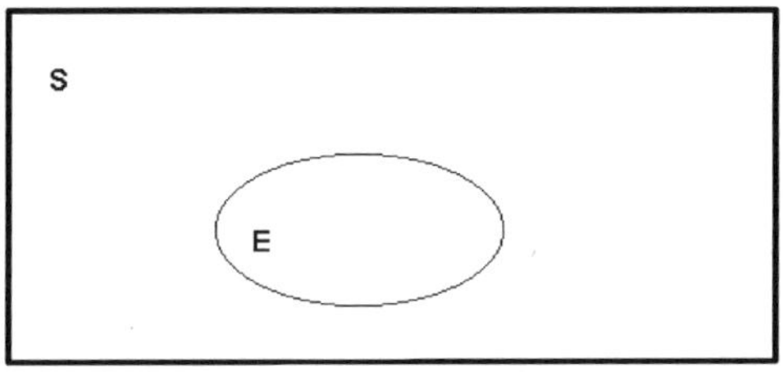

Fig. 4.7 Venn diagram that describes the selected event E and the Sample Space S.

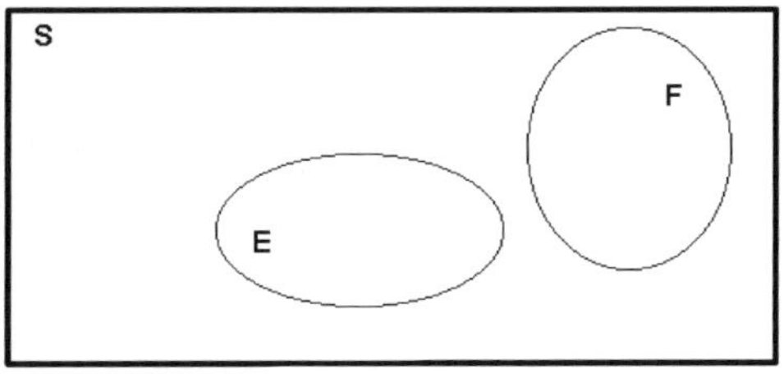

*Fig. 4.8 Venn diagram that describes the sets of incompatible selected events **E** and **F** and of Sample Space **S**.*

In fact, it is easy to calculate that the probability of the Union of the events **E** and **F** is the total of the cardinalities **#E** and **#F**, divided by **#S**:

$$\mathcal{P}(\mathbf{E} \cup \mathbf{F}) = \frac{\#\mathbf{E} + \#\mathbf{F}}{\#\mathbf{S}} = \frac{\#\mathbf{E}}{\#\mathbf{S}} + \frac{\#\mathbf{F}}{\#\mathbf{S}} = \mathcal{P}(\mathbf{E}) + \mathcal{P}(\mathbf{F}) \qquad (4.2)$$

In the case of the third question, mentioned above, the two selected events are not incompatible, so that rule cannot be applied.

We try to determine a more general rule that can apply even when there are complex selected events, compound, for example, by events **E**, **F** and **G** not incompatible (fig.4.9).

We can easily observe from the Venn diagram that the Union of the sets can be constructed as the corresponding area.

If we do only the sum of the cardinalities of **E**, **F** and **G**, we get a higher value, because the cardinalities of overlapping sets, **EF,** for **E** and **F**, **EG,** for **E** and **G,** and **FG** for **F** and **G,** are taken twice, and therefore must be removed once. But in doing so the area **EFG** common to **EF**, **EG** and **FG** is completely eliminated, then it should be re-added.

We can find a formula considering the incompatible parts $\mathbf{E_{Inc.}}$, $\mathbf{F_{Inc}}$, $\mathbf{G_{Inc.}}$, $\mathbf{EF_{Inc.}}$, $\mathbf{EG_{Inc.}}$, $\mathbf{FG_{Inc.}}$, **EFG** of the relevant sets:

$$\#E_{inc} + \#F_{inc} + \#G_{inc} + \#EF_{inc} + \#EG_{inc} + \#FG_{inc} + \#EFG =$$

$$= \#E - \cancel{\#EF} - \cancel{\#EG} + \cancel{\#EFG} + \#F - \#EF - \cancel{\#FG} + \cancel{\#EFG} + \#G - \#EG - \#FG + \cancel{\#EFG} +$$

$$+\cancel{\#EF} - \cancel{\#EFG} + \cancel{\#EG} - \cancel{\#EFG} + \cancel{\#FG} - \cancel{\#EFG} + \#EFG$$

As sum of cardinalities of selected cases we get then:

$$\#E + \#F + \#G - \#EF - \#EG - \#FG + \#EFG \qquad (4.3)$$

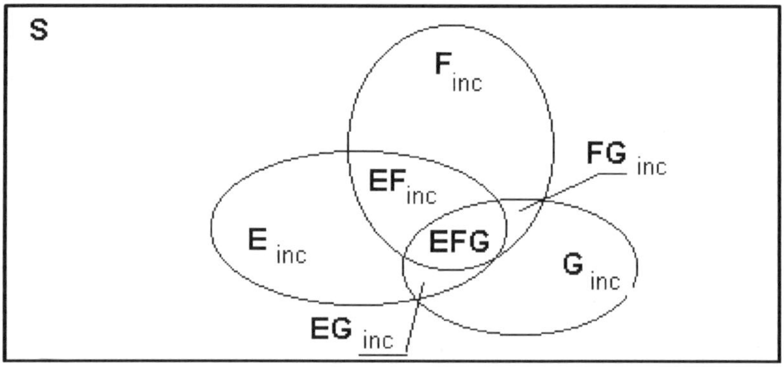

*Fig.4.9 Venn diagram that describes the area of selected events **E**, **F** and **G** not incompatible in the sample space **S**.*

Note that we must add the odd terms (single and triple intersections factors) and subtract even terms (double intersections). This applies in general, being, thus, a rule that can be applied to any number of events, whether they are incompatible or compatible.

We solve now the third problem with this rule. The sample space is 52-4 = 48 cards, which if **E** stands for "suit of clubs" and **F** stands for "a 9" we have **#E** = 13-4 = 9 (9 cards of clubs remain in the deck), **#F** = 4 (four 9 remain in the deck), **#EF** = 1 (the 9 of clubs).

The number of selected cases is then 9+4-1 = 12, and the probability is 12/48, as determined already with the relevant tree diagram.

4.2.4 Conditional Probability

Many of the probability data commonly used have a slightly different meaning from the above described.

Common terms, cited often at discursive level or by newspapers, are, for instance, the chances of recovery from a given disease, the probability of winning at a certain game or other similar. It is evident that the sample space, in this case, is not the whole population.

The probability of recovery from a given disease refers to the subset of the population affected by that disease. If I want to know my risk of dying for that disease, being at present not affected by it, I cannot just make the complement to 1 of the probability of recovery; I shall consider, before the probability to become affected, thereafter the probability not to recover.

The same happens on the technical side. If our Company has a number of machine tools for which their producer has given us the probability of fault, we have to consider before the probability that the given machine is used, thereafter the probability that it has a fault. We see, for example, the following problem:

- In a production department dealing with three types of machine tools $\mathbf{B}_1$, $\mathbf{B}_2$ and $\mathbf{B}_3$, for which the manufacturers claim a relevant probability of failure $\mathcal{P}(\mathbf{D}_1)$, $\mathcal{P}(\mathbf{D}_2)$ and $\mathcal{P}(\mathbf{D}_3)$. What is the global probability of having a fault?

To address this problem, we need to define the concept of conditional probability, $\mathcal{P}(\mathbf{E}|\mathbf{F})$, which represents the probability that event **E** occurs after the event **F** occurred. It is evident that for the user $\mathcal{P}(\mathbf{D}_1)$, $\mathcal{P}(\mathbf{D}_2)$ and $\mathcal{P}(\mathbf{D}_3)$ correspond to this concept, that is the probability of the event "Faulty" (**A**) when the event "Machine in use is of type *i*" ($\mathbf{B}_i$) has happened, a conditional probability with the event $\mathbf{B}_i$ as conditioning event and the event **A** as conditioned event:

$$\mathcal{P}(\mathbf{D}_i) = \mathcal{P}(\mathbf{A} \mid \mathbf{B}_i) \tag{4.4}$$

To assess the probability that the Department will have a failure on a machine of type 1, we can manage the problem with the appropriate tree diagram which takes into account, first, the probability that a machine of type *i* is used $\mathcal{P}(\mathbf{B}_i)$, then that it is faulty $\mathcal{P}(\mathbf{D}_i)$ (fig. 4.10).

To evaluate the conditional probability $\mathcal{P}(\mathbf{E}|\mathbf{F})$ we can use the general definition of probability, considering the set of selected cases, that is the joint happening of **E** and **F**, therefore the intersection of **E** and **F**, as referred to the possible events, and, as we are interested only if the conditioning event has happened, to the set **F.** Referring to fig. 4.11, we see that the selected cases are given by the cardinality **#EF**, while possible cases by the cardinality **#F**:

$$\mathcal{P}(\mathbf{E}|\mathbf{F}) = \frac{\#\mathbf{EF}}{\#\mathbf{F}} = \frac{\frac{\#\mathbf{EF}}{\#\mathbf{S}}}{\frac{\#\mathbf{F}}{\#\mathbf{S}}} = \frac{\mathcal{P}(\mathbf{EF})}{\mathcal{P}(\mathbf{F})} \tag{4.5}$$

We can also see that:

$$\mathcal{P}(\mathbf{EF}) = \mathcal{P}(\mathbf{F}) \cdot \mathcal{P}(\mathbf{E}|\mathbf{F}) \tag{4.6}$$

which tells us that the probability to have together the events **E** and **F** (in the Sample Space **S**) is given by the product of the probability of that happening **F** (conditioning event) times the conditional probability that occur **E** when **F** has already happened, as represented by the tree diagram.

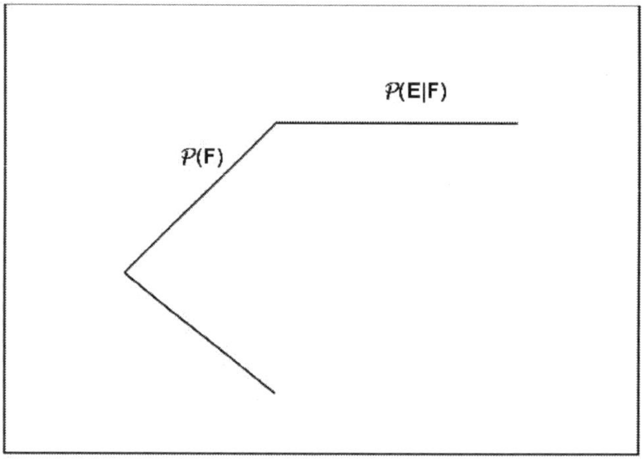

Fig. 4.10 Tree diagram showing the meaning of conditional probability.

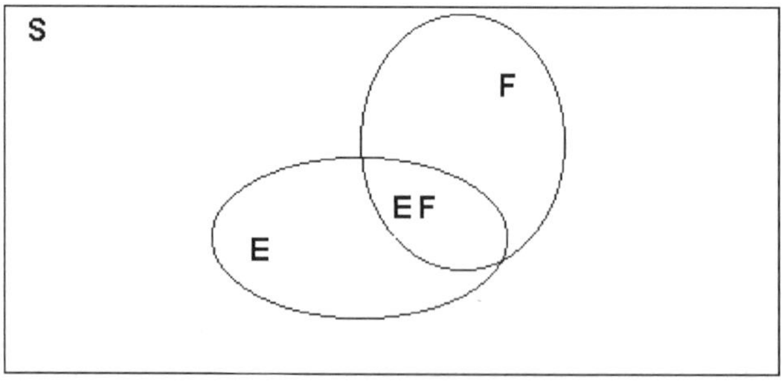

*Fig.4.11 Venn diagram that describes, for the conditional probability $\mathcal{P}(\mathbf{E}|\mathbf{F})$, the area of **EF,** favourable events, and the corresponding sample space **F** (not **S**).*

A further condition often is to take into account all possible alternatives, i.e., the situation of the whole production department. The use of the different machines is represented by sets $\mathbf{B}_1$, $\mathbf{B}_2$ and $\mathbf{B}_3$**.** That sets are incompatible but also comprehensive, that means that no other machines are used in the department, as seen in figure 4.12.

In that case, the Sample Space **S** is the union of $\mathbf{B}_1$, $\mathbf{B}_2$ and $\mathbf{B}_3$. We can therefore easily calculate the probability *P* ($\mathbf{B}_i$) of use of a machine of the ith type:

$$\mathcal{P}(\mathbf{B}_i) = \frac{\#\mathbf{B}_i}{\#\mathbf{B}_1 + \#\mathbf{B}_2 + \#\mathbf{B}_3} \tag{4.7}$$

The tree diagram that evaluates the general probability of faults is shown in fig. 4.13

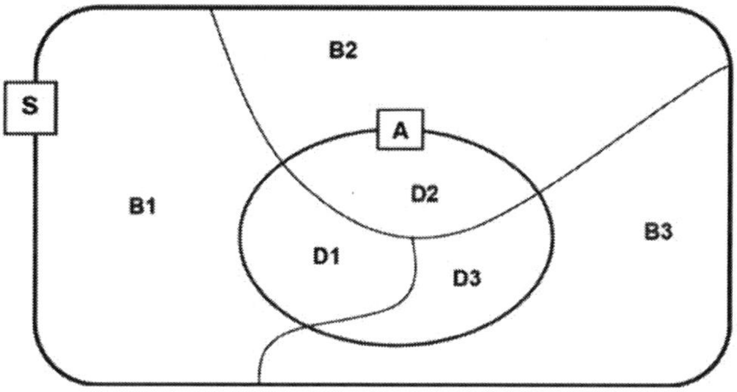

Fig. 4.12 Venn diagram that describes the sets for the three groups of machines $\mathbf{B}_1$, $\mathbf{B}_2$ and $\mathbf{B}_3$, for their faulty conditions $\mathbf{D}_1$, $\mathbf{D}_2$ and $\mathbf{D}_3$, for any type of fault **A**.

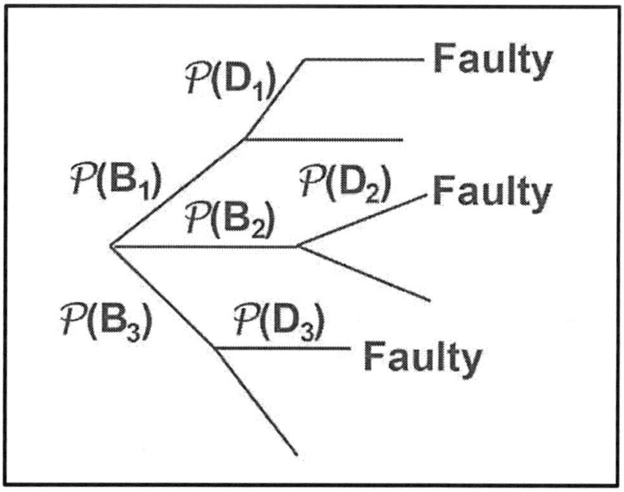

Fig. 4.13 Tree diagram for the evaluation of the probability of faults

As we can see the total probability $\mathcal{P}(\mathbf{A})$ of faults is the sum of the probability that we will get following three different branches:

$$\mathcal{P}(\mathbf{A}) = \sum_i \mathcal{P}(\mathbf{B}_i) \cdot \mathcal{P}(\mathbf{D}_i) = \sum_i \mathcal{P}(\mathbf{B}_i) \cdot \mathcal{P}(\mathbf{A} \mid \mathbf{B}_i) \tag{4.8}$$

This formula is called "Total Probability Theorem" and can be used to solve another problem, that is, to analyse the reverse path, which, for the example given, corresponds to determine when a fault has occurred, what is the probability that it is due to a machine of the ith type. We want, therefore, determine the conditional probability $\mathcal{P}(\mathbf{B}_i|\mathbf{A})$.

We can write, referring to fig. 4.13, and using the definition of conditional probability:

$$\mathcal{P}(\mathbf{B}_i \mid \mathbf{A}) = \frac{\mathcal{P}(\mathbf{B}_i\mathbf{A})}{\mathcal{P}(\mathbf{A})} \tag{4.9}$$

And clearly being $\mathcal{P}(\mathbf{B}_i\mathbf{A}) = \mathcal{P}(\mathbf{A}\ \mathbf{B}_i) = \mathcal{P}(\mathbf{B}_i)\cdot\ \mathcal{P}(\mathbf{A}|\mathbf{B}_i)$, we have:

$$\mathcal{P}(\mathbf{B}_i|\mathbf{A}) = \frac{\mathcal{P}(\mathbf{B}_i)\cdot\mathcal{P}(\mathbf{A}\mid\mathbf{B}_i)}{\mathcal{P}(\mathbf{A})} = \frac{\mathcal{P}(\mathbf{B}_i)\cdot\mathcal{P}(\mathbf{A}\mid\mathbf{B}_i)}{\sum_{k}\mathcal{P}(\mathbf{B}_k)\cdot\mathcal{P}(\mathbf{A}\mid\mathbf{B}_k)} \tag{4.10}$$

called Bayes' formula.

5 BEYOND DETERMINISM: NOTES ON STATISTICS

5.1 STATISTICAL DISTRIBUTIONS

Often, instead of a particular aspect, concerns are about the global landscape for a given situation. For example, if we look at the analysis of the results of the launch of two dice, we may be interested in knowing the probability of having a sum equal to 8, but most commonly we are interested in learning about the global trend of probabilities in order to know if there is a condition more likely to bet on and what advantage this condition may result than the other. If we look at the situation as a whole, we can describe a trend that represents the distribution of outcomes, which can be transformed into a probability distribution (Fig. 4.2 shown here in Fig. 5.1, for the reader's convenience).

						6-1					
					5-1	5-2	6-2				
				4-1	4-2	4-3	5-3	6-3			
			3-1	3-2	3-3	3-4	4-4	5-4	6-4		
		2-1	2-2	2-3	2-4	2-5	3-5	4-5	5-5	6-5	
Outcomes	1-1	1-2	1-3	1-4	1-5	1-6	2-6	3-6	4-6	5-6	6-6
Sum	2	3	4	5	6	7	8	9	10	11	12

Fig. 5.1 Scheme for the probability of the sum of outcomes rolling two dice.

In the case of the two dice, it can be said, from the shape, that we have a triangular probability distribution.

Note that, for representing a statistical distribution, it is necessary to put the values of the variable considered (in the example the sum of the results of two outcomes) on a Cartesian axis, usually along the abscissa axis, while on the other axis an indication of the probability is given. Statistical distributions can be obtained theoretically, considering general characteristics of the phenomenon examined, or experimentally. This situation is very similar to that of classical and frequentistic definition of probability. From the theoretical point of view the distribution can be perfectly defined, but a "correlation uncertainty" can arise from disagreement between theoretical hypotheses and the real phenomenon (the definition of uniform distribution for the outcomes of a dice is based on conditions of symmetry and uniformity; if the dice is not symmetrical or its material not uniform, the distribution will deviate from the uniform one). From the experimental point of view, as seen for the frequentistic definition of probability, it is necessary to have many data (of the order of a thousand) to obtain a robust definition, that is a definition able to be confirmed with small changes in case the experiments were repeated. In practice, it is very rare that we can have the number of experimental data needed to completely describe a statistical distribution on the experimental basis. Therefore the usual way is to examine the characteristics of the phenomenon considered to connect them in the best possible way to theoretical hypotheses depicting a corresponding theoretical statistical distribution. In fact, in principle, a theoretical distribution can be represented by a mathematical function with some parameters (e.g. the equation of a line $y = a + bx$ with the parameters a and b). The mathematical function can be identified by evaluating the parameters with a number of experimental data not so large as necessary for evaluating the whole shape of the distribution (as it is done to find the line passing between two experimental points whose coordinates are used to calculate the values of a and b). For example a distribution not very common, but very useful for the management of experimental data, is the *U* distribution, described by the equation $[\pi(a^2-\theta^2)^{1/2}]^{-1}$. Its name comes from its shape: when the variable θ assumes values $+a$ or $-a$ the function tends to infinity, while for $\theta = 0$ assumes the minimum value, therefore it has the shape of a *U* with the extreme parts of the right and left high and the central part low. In this case the experimental data shall identify the $2a$ interval, that must be identified with about 100 data, quite high, but frequently practicable.

The complete description of the statistical distribution allows fully exploiting the statistical information for very delicate activities; however frequently a limited use of statistics is sufficient for obtaining good advantages.

For example, in the practice of the management of experimental data, we have frequently groups of results for each variable, but for our calculation we need a single value. Consider we measured the sides b and h of a sample for traction test of rectangular section, as we need the area $A = bh$ for calculations. What interests us is to extract from the groups of results obtained for b and h the values that best represent the lengths of the sides. With statistical terminology, we want to evaluate the *expected values* of the statistical variables B and H associated to the measured values b and h (note which care is used, in the field of statistics, to distinguish, even in the symbolism, the variable, indicated as symbol in uppercase, and the corresponding values, indicated in lowercase).

Often, in everyday life to have the expected value, calculated just using the average, is considered a satisfying situation, and also in technical applications is frequently so. But this is the statistic mocked by Trilussa, an Italian poet who says: "*if we are two and there are two roast chickens, for statistics we eat one each, even if you've eaten them both!*" It does not work, but often this is the current application, even if statistics formally requires to evaluate always the variability. As was shown in Chapter 1, we always need to associate to any information the relevant value and its variability. At least these two data are necessary, allowing us to manage, albeit with some approximation, the information obtained experimentally.

It is possible to work even if we do not know the shape of the statistical distribution, but we need, at least, to know in which position of the axis of the variable the distribution is set, which is given by the expected value of the variable, that is called also *position index*, and an expected value of variability, that is called *dispersion index*. To allow a less approximate evaluation other two indexes are used, one indicating how the distribution is symmetrical, called *symmetry*, and the other whether it is flat or sharp, called *kurtosis*. But the number of data required for a robust evaluation of these indexes is fairly high, therefore they are seldom used.

To proceed, we need now to go a little more deep into the concepts connected with statistical distributions.

5.2 MAIN TYPES OF STATISTICAL DISTRIBUTIONS

Speaking about probability, we used some random phenomena, such as the roll of a dice, that help illustrating with events well known to all what may be the relevant probability. So, for a dice with six sides, there is a given probability of getting the side 1, or 2, ..., or 6. The output is clearly discontinuous; the event side 1.2 does not exist. The typology of the phenomenon thus described also covers a number of technical applications.

If, when testing of industrial production, I associate my assessment of the number of parts out of tolerance, I can find as result only an integer, 0, or 1, or 2, ..., k pieces out of tolerance. The event "1.2 pieces out of tolerance" does not exist.

If, however, we consider a situation of a very high number of workpieces, for example the production of millions of objects, with the presence of a certain feature in 10% of the specimens, we would be dealing with values such as 100 000, 100 001, 100 002.

They are certainly discontinuous, but the discontinuity phenomena referred to the number almost disappears, so much that we would prefer to give the result as a percentage and consider the variation of percentage virtually continuous. Moreover, it must be remembered that some characteristics of the product, for example the dimensions, are inherently continuous, so if we want to evaluate our production process of cylindrical pins, with a defined tolerance on diameter and cylindricity, we must consider the distributions of continuous variables.

A first subdivision, therefore, must be pointed out between the statistical distributions of discrete variables (with values discontinuous), as the number of pieces out of tolerance in a production batch not numerous, or continuous[1] variables, such as the geometric dimension of a measurand (diameter of a cylindrical pin). These two cases are handled differently as graphical presentation and as calculation.

In the case of discrete distributions there is a number of defined, discontinuous values of the variable, to which corresponds a certain probability. Therefore the graphic representation can be made by placing these values of

---

[1] Generally phenomena are in most cases discontinuous (taking into account the measurement resolution, even continuous quantities are managed in a discontinuous way). In any case a continuous variable is useful when the discontinuity is negligible for the considered events (e.g. in case of measurements or evaluation of large batches, frequently expressed in percentage).

the variable on the abscissa and the corresponding values of probability in ordinate. As examples, we can see in Fig. 5.2 the representations of the uniform distribution corresponding to the outcomes of a six-sided dice and in Fig. 5.3 the triangular distribution of the sum of the outcomes obtained rolling two six-sided dice.

In most of the cases related to experimental data a counting or a measurement is involved: in both cases the information obtained is discontinuous (for measurement think about reading resolution). But, as it is usually done presenting the measurement results in a continuous way, also for the statistical distribution the continuous condition is preferred. For the presentation of measurement results the usual way is to fill the spaces between subsequent discontinuous values with a constant continuous value.

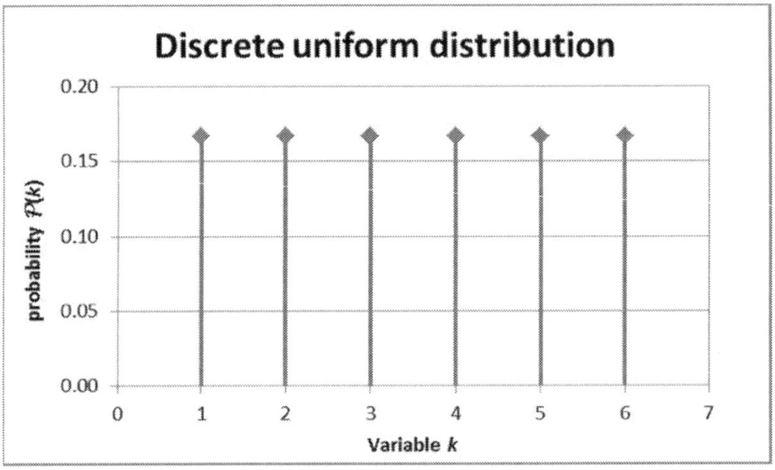

Fig. 5.2 Statistical distribution of the outcomes obtained rolling one dice (uniform discrete distribution).

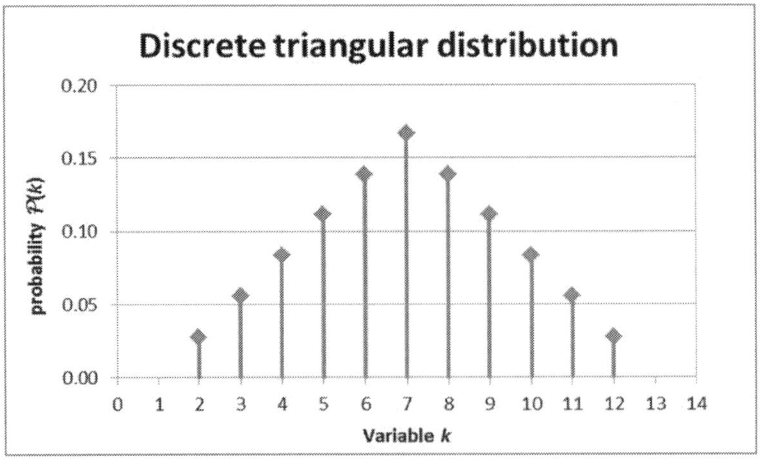

Fig. 5.3 Statistical distribution of the sum of the two outcomes obtained rolling two dice (triangular discrete distribution).

Let us try the same for our triangular distribution: we will get the bar chart of Fig. 5.4, but the point is that the ordinate value cannot be the probability, as the continuous representation gives ordinate values for each point of the abscissa. Now we can consider the probability of getting a result in a given interval of values of our variable, but if we reduce the interval to a really infinitesimal dimension, as a point, the relevant probability shall be infinitesimal. For this reason Karl Pearson introduced the graphical presentation called *histogram*. Pearson noted, as can be seen from the given example which has the bar bases equal to 1 and their height equal to the relevant probability, that probability can be represented by the area of bars (and this is in line with the fact the probability on a single point is null).

This is, in fact, the special characteristic of the histogram, which represents probability with the area, not with the height of the bars.

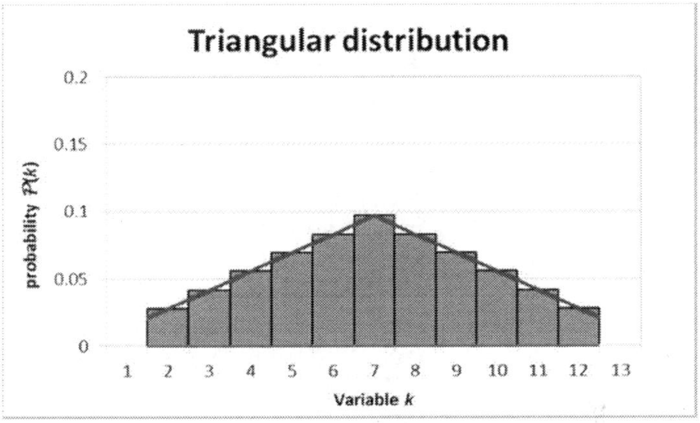

Fig. 5.4 Transformation of the discrete statistical distribution[2] of the sum of two outcomes obtained rolling two dice in a continuous one passing through a bar chart.

Pearson had, also, to consider the meaning of the heights, that is of the ordinate of the graph, which, resulting to be the area divided by the base of each rectangle, that is the probability divided by the relevant interval of the variable, was called *density of probability*.

So, if we have a graph representing the statistical distribution of the continuous variable x and we want to know the probability that x is between a and b, we shall evaluate the area of the graph in the interval a to b. If the density of probability is described by the function $f(x)$, we shall evaluate the integral given in equation 5.1:

$$F(b) - F(a) = \int_a^b f(x)dx \qquad (5.1)$$

[2] Note the discontinuity on the vertex. For evaluating the probability in the interval of the vertex one can use the density function on the left up to x=3.75 or on the right starting from x=3.25.

Check that for our triangular distribution the probability that x is between 2.5 and 3.5 is equal to the probability of the sum of outcomes equal to 3 for two six-sided dice, 3/36. In this case the density of probability is given in equation 5.2 together with the integral calculation that shows what required:

$$\begin{aligned} &\mathrm{f}(x)=\frac{x}{36} \\ &F(b)-F(a)=\int_a^b \mathrm{f}(x)dx=\frac{b^2-a^2}{2\cdot 36}=\frac{(3+0.5)^2-(3-0.5)^2}{2\cdot 36}= \\ &=\frac{3^2+2\cdot 3\cdot 0.5+0.5^2-3^2+2\cdot 3\cdot 0.5-0.5^2}{2\cdot 36}=\frac{4\cdot 3\cdot 0.5}{2\cdot 36}=\frac{3}{36} \end{aligned} \tag{5.2}$$

Note that the x variable can change, but the statistical distribution based on the density of probability remains consistent. For instance, in the same process of two six-sided dice we can consider as variable x the average of the two outcomes: in this case possible results are 1, 1.5, 2, ..., 6, with the same shape as the distribution of the sum, but the histogram and the relevant continuous distribution change I such a way to keep the same probability for the interval corresponding to each discrete value. So the probability for the discrete outcome 1.5 is again 3/72, and we can check from Fig. 5.5 that the integral on the corresponding interval between 1.25 and 1.75 is the same value 3/72.

$$\begin{aligned} &\mathrm{f}(x)=\frac{x}{18} \\ &F(b)-F(a)=\int_a^b \mathrm{f}(x)dx=\frac{b^2-a^2}{2\cdot 18}=\frac{(1.5+0.25)^2-(1.5-0.25)^2}{2\cdot 18}= \\ &=\frac{1.5^2+2\cdot 1.5\cdot 0.25+0.25^2-1.5^2+2\cdot 1.5\cdot 0.25-0.5^2}{2\cdot 18}=\frac{4\cdot 1.5\cdot 0.25}{2\cdot 18}=\frac{3}{72} \end{aligned} \tag{5.3}$$

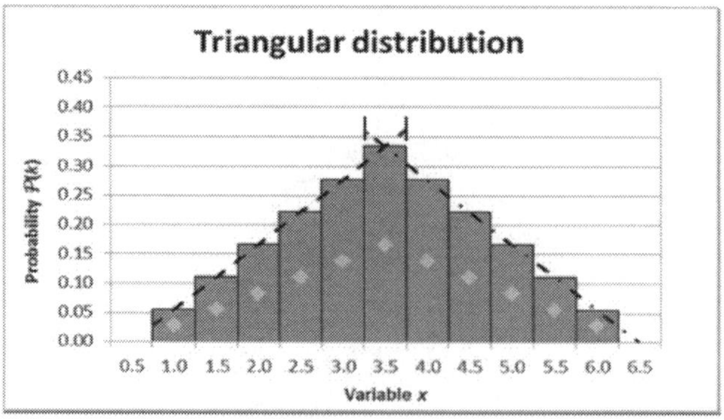

Fig. 5.5 Transformation of the discrete statistical distribution of the average of two outcomes obtained rolling two dice in a continuous one passing through a bar chart. Note how the histogram representation (bar area) is consistent with the discrete distribution: as the bar base is 0.5, the height is doubled, so that the bar area is equal to the discrete probability.

Summarizing it is important to remember that it can be useful to evaluate phenomena through different types of statistical distributions, which can be discrete or continuous as better suited for the type of statistical variable considered. For discrete distribution for each discontinuous value the relevant probability is given, while for continuous distribution the probability can only be associated with an interval from a to b of the statistical variable x, and the probability will be obtained considering the function density of probability $f(x)$ by the integral given in equation 5.1.

5.3 MAIN DISTRIBUTION PARAMETERS

As said before the main points of interest are:

- which value can I associate to a group of results (or, more in general, to a statistical variable);
- what is the variability that I shall associate to my results (or, more in general, to a statistical variable).

The statistical answer to these questions is obtained using the concept of the expected value.

5.3.1 The position index

If we have a group of experimental results and we will evaluate a single value to use in calculation, we face a problem similar to that of having a group masses to be substitute by a single mass having the same effect. We know that we have to calculate the centre of mass position. Our case is very similar and we use equations similar to that applied for the cases of a group of discrete masses or of a single, continuous mass having a given shape.

If we have a statistical variable X, the expected value of X gives us a sort of point of equilibrium, so that the position index is frequently called *central value*, but this terminology shall be used carefully because the position index corresponds to the central value only for symmetrical distributions.

More frequently it is called the average, as it balance the effects of data around it and, moreover, as we will see later, because it can be evaluated by the arithmetic average.

As you know there are different averages, as geometric average, harmonic average, but generally the simple use of the term “average” means the arithmetic average. It is very important to state immediately a fundamental distinction from theoretical average and experimental average.

If we consider a theoretical distribution, e.g. the distribution of outcomes of a six-sided dice, we evaluate the arithmetic average on theoretical data and we get a theoretical average that is indicated by the symbol μ: the value obtained is an exact value. If we evaluate the average outcome of our dice (exactly that dice) rolling it n times, we can do the average of the outcomes and divide it by n. In that case we get an estimation of the average of our dice. If we get other n outcomes, we certainly obtain a different value for the average.

The result obtained has a variability, therefore is a statistical variable frequently indicated by the symbol m (from the word mean) or by the symbol of the value of the variable averaged overwritten with a bar (variable X, value x, mean value $\bar{x}$).

Expected value of a discrete distribution

Let us consider a statistical variable X which can attain discontinuous values x_{si} each one with the relevant probability $\mathcal{P}(x_{si})$. The expected value of X is defined by equation 5.4. This can be done using the classical definition of probability or the frequentistic definition of probability.

If we consider a perfect six-sided dice we can use the classical definition: each possible outcome x_{si}, 1, 2, ..., 6, has an equal probability $\mathcal{P}(x_{si})$=1/6. The expected value is therefore given in equation 5.5.

$$\mathrm{E}(X)=\sum_{i=1}^{k}\mathcal{P}(x_{si})\cdot x_{si} \tag{5.4}$$

$$\mathrm{E}(X)=\sum_{i=1}^{k}\mathcal{P}(x_{si})\cdot x_{si}=\sum_{i=1}^{6}\frac{i}{6}=\frac{1+2+3+4+5+6}{6}=3.5 \tag{5.5}$$

If we shall manage experimental data, we shall use frequentistic definition. For instance if we measured the sides b and h of a sample for traction test of rectangular section, to calculate the area $A = bh$, we shall get from our data the expected values of the variables B and H representing the lengths of the sides.

Usually we need to characterize the measurement process, that is, at the first application of the process we do a lot of measurements to understand how results can be statistically distributed, to decide if the capacity of our process is sufficient and how many measurements should be necessary to be in a robust condition. To give you an example, 100 results are reported in Table 5.1, for values b, and in Table 5.2, for values h.

We work now on the values b obtained for the variable B. Using the frequentist definition of probability we can evaluate, as shown in Table 5.3, the discontinuous distribution of our data.

*Tab. 5.1 Experimental results of the measurement of the variable **B**. Data are expressed in millimetres and, as can be seen, are discontinuous with a resolution (discontinuity) of 0.01 mm.*

12.09	11.90	12.10	12.09	12.00	12.07	12.07	11.99	12.09	12.04
12.06	12.08	11.99	11.91	11.98	12.02	12.08	12.05	12.07	11.99
11.90	11.99	11.90	12.05	11.96	11.99	11.94	11.97	11.91	12.03
11.91	12.08	12.07	11.92	11.95	11.97	12.10	12.02	11.96	12.04
12.06	12.09	11.98	11.91	12.08	11.93	11.92	12.00	11.91	11.92
11.99	11.90	11.97	12.10	11.94	12.05	12.06	12.00	11.97	11.93
11.92	11.92	12.00	12.03	11.92	12.08	11.97	12.02	12.09	11.92
11.95	12.06	11.90	12.00	12.00	11.93	12.04	11.94	12.10	12.02
12.10	12.10	12.02	12.01	11.95	12.03	12.00	12.08	11.94	11.98
12.08	11.94	12.01	11.96	11.99	11.99	11.90	11.96	12.04	11.93

*Tab. 5.2 Experimental results of the measurement of variable **H**. Data are expressed in millimetres and, as can be seen, are discontinuous with a resolution (discontinuity) of 0.01 mm.*

7.09	7.04	6.94	7.09	7.01	6.93	7.05	6.90	6.99	6.97
7.01	6.93	7.05	6.94	6.95	7.00	7.07	6.95	6.97	6.92
6.92	6.99	7.02	6.90	6.99	6.99	6.98	6.93	7.09	7.03
6.90	7.00	6.91	7.00	7.07	7.03	7.03	7.03	6.92	7.06
7.07	6.93	6.93	7.03	7.02	7.09	6.94	7.02	6.97	6.93
6.95	6.96	7.06	6.92	7.02	6.98	7.08	7.10	7.08	7.01
7.05	6.92	7.03	6.94	6.95	6.98	6.92	7.05	6.99	7.08
6.96	7.09	7.04	6.91	6.99	6.95	6.94	6.90	6.90	7.01
6.93	6.98	7.01	6.97	7.01	6.90	7.04	6.97	6.97	7.03
6.92	7.05	6.96	6.95	6.95	7.06	6.97	7.08	7.06	7.02

Tab. 5.3 For each value b_i is evaluated the relevant probability $\mathcal{P}(\mathbf{b}_{si})$. Counting the number f_{ai} of results having $\mathbf{b}_{si}$ value (f_{ai} is called absolute frequency) and dividing it by the total number of data we obtain the relative frequency f_{ri} which is the experimental evaluation of $\mathcal{P}(\mathbf{b}_{si})$.

$\mathbf{b}_{si}$/mm	f_{ai}	$\mathcal{P}(b_{si})$
11.90	6	6.0%
11.91	5	5.0%
11.92	7	7.0%
11.93	4	4.0%
11.94	5	5.0%
11.95	3	3.0%
11.96	4	4.0%
11.97	5	5.0%
11.98	3	3.0%
11.99	8	8.0%
12.00	7	7.0%

$\mathbf{b}_{si}$/mm	f_{ai}	$\mathcal{P}(b_{si})$
12.01	2	2.0%
12.02	5	5.0%
12.03	3	3.0%
12.04	4	4.0%
12.05	3	3.0%
12.06	4	4.0%
12.07	4	4.0%
12.08	7	7.0%
12.09	5	5.0%
12.10	6	6.0%

For each possible state of the discontinuous variable B (in our example 11.90, 11.91, 11.92. .... 10.10; note the gap of 0.01 mm between the k=21 different states over n=100 data) we count the number of data (absolute frequency f_{ai}), then we calculate, dividing it by the total data number n=100, the relative frequency f_{ri}, which is an estimation of the probability $\mathcal{P}(b_{si})$ that a result get the value b_{si}. Having the values $\mathbf{b}_{si}$ of the different states and relevant probability $\mathcal{P}(b_{si})$ we can calculate the expected value of B with equation 5.6:

$$\mathrm{E}(B)=\sum_{i=1}^{k}\mathcal{P}(b_{si})\cdot b_{si} \tag{5.6}$$

As a general note we can observe that summation is not done over the number n of data, but on the number k of different states the variable X can attain. Note also that if we substitute the absolute frequency for calculating $\mathcal{P}(x_{si})$, we get:

$$\mathrm{E}(X)=\sum_{i=1}^{k}\mathcal{P}(x_{si})\cdot x_{si}\approx\sum_{i=1}^{k}\frac{f_{ai}}{n}x_{si}=\sum_{j=1}^{n}\frac{x_j}{n} \tag{5.7}$$

that is, we can consider that the summation (over i, indicating the ith state if the variable X) up to k (number of different states of X) of the number f_{ai} of data in the ith state is the total number n of data (j in the last part indicate one of the n data). Therefore the rightmost member of the equation 5.7 indicates that the expected value can be estimated by the arithmetical average of data. You can yourself to estimate the expected value and the average for variable H.

Expected value of a continuous distribution

As usual the change from discrete functions to continuous functions involves going from summations to integrals, and indeed the formula for expected value change in equation 5.8:

$$\mathrm{E}(X) = \int_{-\infty}^{\infty} x \cdot \mathrm{f}(x) dx \tag{5.8}$$

We can apply this formula to the uniform continuous distribution that corresponds to the outcomes of a six-sided dice. Each outcome involves, a unit interval (getting 2 involves the interval from 1.5 to 2.5, therefore we assign to 1 the interval between 0.5 to 1.5 and to 6 the interval 5.5 to 6.5). Being 1/6 the probability of each outcome, our histogram has bars wide 1 and height 1/6. A good check always to do is that total area shall be equal to 1, as it is here. Therefore in our case we have:

$$\mathrm{E}(X) = \int_{-\infty}^{\infty} x \cdot \mathrm{f}(x) dx = \int_{0.5}^{6.5} x \cdot \frac{1}{6} dx = \left. \frac{x^2}{2 \cdot 6} \right|_{0.5}^{6.5} = \frac{42.25 - 0.25}{12} = 3.5 \tag{5.9}$$

Check yourself what is the expected value of the continuous triangular distribution corresponding to the sum of outcomes of two six-sided dice. You should get 7.

Some properties of the expected value and of the average

It is useful to evaluate the expected value of the sum and of the product of variables. Here we use the formulas for discrete distributions, but it is easy to understand that these properties apply also to continuous distributions.

For the sum is more easy to use the definition based on the average, as seen in 5.7, here rewritten in 5.10 for the variable Y which is the sum of g variables X_l:

$$\begin{aligned} Y &= \sum_{l=1}^{g} X_l \\ \mathrm{E}(Y) &= \sum_{j=1}^{n} \frac{y_j}{n} \\ \mathrm{E}\left(\sum_{l=1}^{g} X_l\right) &= \frac{1}{n} \sum_{j=1}^{n} \sum_{l=1}^{g} x_{lj} = \sum_{l=1}^{g} \sum_{j=1}^{n} \frac{x_{lj}}{n} = \sum_{l=1}^{g} \mathrm{E}(X_l) \end{aligned} \tag{5.10}$$

We can see that the expected value of the sum is equal to the sum of expected values.

For the product we use of the definition of the expected value given in 5.4. Before all we shall see the structure of the different states of Z=XY. Consider, for instance, that you have X on 2 different states, x_{s1} and x_{s2}, and Y on 3 different states, y_{s1}, y_{s2} and y_{s3}, the states of Z are $x_{s1}y_{s1}$, $x_{s1}y_{s2}$, $x_{s1}y_{s3}$, $x_{s2}y_{s1}$, $x_{s2}y_{s2}$, $x_{s2}y_{s3}$. We have therefore:

$$Z = X \cdot Y$$

$$\mathrm{E}(Z) = \sum \mathcal{P}(z_{si}) \cdot z_{si} = \sum_{i=1}^{k} \sum_{j=1}^{l} \mathcal{P}(x_{si} \cdot y_{sj}) \cdot x_{si} \cdot y_{sj}$$

There is a simple evolution if the variables X and Y are independent, in fact in that case the probability of the product is equal to the product of probabilities and we have:

$$\mathrm{E}(X \cdot Y) = \sum_{i=1}^{k} \sum_{j=1}^{l} \mathcal{P}(x_{si}) \cdot \mathcal{P}(y_{sj}) \cdot x_{si} \cdot y_{sj} = \sum_{i=1}^{k} \mathcal{P}(x_{si}) \cdot x_{si} \cdot \sum_{j=1}^{l} \mathcal{P}(y_{sj}) \cdot y_{sj} = E(X) \cdot E(Y) \quad (5.11)$$

Therefore, in case of independent variables the expected value of the product is equal to the product of expected values.

One important point is the relationship between theoretical and experimental values of the position index, μ and m.

As said before they are different in value but also as mathematical entities: for a given distribution μ is a constant, while m, obtained from experimental data, is an estimation of μ that changes when the set of experimental data is different, therefore m is a statistical variable.

We can highlight some properties of m, which can be calculated as arithmetical average, following the rightmost part of equation 5.7. First of all we can show that the sum of the residuals from m (the differences of experimental data x_j from m) is identically null:

$$\sum_{j=1}^{n} (x_j - m) = \sum_{j=1}^{n} x_j - nm = \sum_{j=1}^{n} x_j - n \sum_{j=1}^{n} \frac{x_j}{n} = \sum_{j=1}^{n} x_j - \sum_{j=1}^{n} x_j = 0 \quad (5.12)$$

this property is very important and will be used frequently.

A second important property refers to the type of estimation of μ we get from m. Statistically it is possible to have correct estimators or incorrect estimators. For correct estimators the differences from the estimated quantity are random, while for incorrect estimator there is a systematic difference. If one estimator is correct, its expected value shall be equal to the estimated quantity, therefore we have to evaluate if E(m)=μ:

$$E(m) = E(\sum_{j=1}^{n} \frac{x_j}{n}) = \sum_{j=1}^{n} \frac{E(x_j)}{n} = \sum_{j=1}^{n} \frac{\mu}{n} = \mu \quad (5.13)$$

therefore we can conclude that m is a correct estimator of μ.

5.3.2 The dispersion index

The second question we posed before is that we want to know the interval in which our data can fall, that is if our distribution is sharp, all data near to the position index (that we have seen is equivalent to the average), or dispersed, large distances among different data.

We shall, therefore, have something connected with the distance among data or more efficiently with the distance of data from the average. But consider as usual the outcomes of a six-sided dice. The average is 3.5, so we can note that the value 6 has a difference 2.5, but the value 1 a difference -2.5, therefore we cannot use the simple difference, because we are interested to the distance from average, which is equal for the value 1 and for the value 6.

We can resort to the square of the difference (Euler's norm) using the sum of squares as index and looking to conditions that reduce to a minimum the sum of squares (frequently called least squares method) or to the absolute value of the difference (Chebyshev's norm), using the sum of absolute values of the differences as index and minimizing it (frequently called Minimax method).

We consider now a procedure based on the sum of squares of the differences; in this case the main dispersion index is called *variance*.

Here too it is very important to state the fundamental distinction from theoretical dispersion index and experimental dispersion index.

If we consider a theoretical distribution we evaluate the variance on theoretical data and we get a theoretical exact value that is indicated by the symbol σ^2.

If we evaluate the variance from experimental data, we get an estimation of the average of our dice. The result obtained has a variability, therefore it is a statistical variable frequently indicated by the symbol s^2. Variance is directly relied to the sum of squares, therefore its dimension is the square of the dimension of the relevant variable, therefore variance is not of immediate use for presenting variability (e.g. if you get a result x=10 mm with its variance of 0.01 mm^2, you do not easily associate a variability related to 0.1 mm to your result), therefore, for indicating variability, it is frequently preferred to use the *standard deviation* which is the square root of the variance and it is indicated by the symbol σ or s, respectively if evaluated theoretically or experimentally.

The dispersion index is evaluated calculating the expected value of the square of the differences between data and the theoretical average μ, as given in equation 5.14.

$$\mathrm{E}(X-\mu)^2 \tag{5.14}$$

this has different forms if applied to discrete distributions or continuous distributions. Application to a discrete distribution of equation 5.14 gives the variance as indicated in equation 5.15:

$$\sigma^2 = \mathrm{E}(X-\mu)^2 = \sum_{i=1}^{k}(x_{si}-\mu)^2 \mathcal{P}(x_{si}) = \sum_{i=1}^{k}(x_{si}-\mu)^2 \frac{\mathrm{f}_{ai}}{n} = \sum_{j=1}^{n}\frac{(x_j-\mu)^2}{n} \tag{5.15}$$

while to continuous distributions, gives what is seen in equation 5.16:

$$\sigma^2 = \mathrm{E}(X-\mu)^2 = \int_{-\infty}^{\infty}(x-\mu)^2 \cdot \mathrm{f}(x)dx \tag{5.16}$$

These calculations can be done with theoretical distributions: in that case it is possible to calculate μ and, therefore, to use formulas 5.15 or 5.16 given before. But in the case of experimental data, we have m and we can calculate s^2; this cannot be done just substituting m to μ in 5.15. To see what to do, let us evaluate, before, if we have an experimental distribution with a variance σ^2 and an average m of q data (we use q to underline that it is the number of data averaged, because frequently one has a total number n of data, but the average is done only on the part q of them, and one common error, when the formula is given using n, is to use the total number of data in calculations), what is the variance σ_m^2 of m:

$$\sigma_m^2 = \mathrm{E}(m-\mu)^2 = \mathrm{E}\left(\frac{\sum_{j=1}^{q} x_j}{q} - \mu\right)^2 = \frac{1}{q^2}\mathrm{E}\left(\sum_{j=1}^{q} x_j - q\mu\right)^2$$

But considering that $q \cdot \mu$ is the average of the sum of q data and the definition of variance, $\mathrm{E}\left(\sum_{j=1}^{q} x_i - q\mu\right)^2$ is the variance of the sum of q equal variables X_i, as seen in 5.10, to the sum of variances of X_i, but being all data used to calculate m taken from the same distribution, their variances are all equal to σ^2, therefore we have:

$$\sigma_m^2 = \frac{1}{q^2}\sum_{i=1}^{q}\sigma^2 = \frac{1}{q^2}q\sigma^2 = \frac{\sigma^2}{q} \tag{5.17}$$

This formula applies approximately also to the variance estimation s^2 calculated using m, as shown hereafter.

We consider, now, n data x_j their differences from μ. We introduce in the formula the average m of the n data adding and subtracting it, so that we have the terms of the differences of the experimental values from m and the difference of m from μ:

$$\sum_{j=1}^{n}(x_j-\mu)^2 = \sum_{j=1}^{n}[(x_j-m)+(m-\mu)]^2 =$$

$$= \sum_{j=1}^{n}\left[(x_j-m)^2 + 2(x_j-m)(m-\mu) + (m-\mu)^2\right] =$$

$$= \sum_{j=1}^{n}(x_j-m)^2 + 2(m-\mu)\sum_{j=1}^{n}(x_j-m) + n(m-\mu)^2$$

Remembering, now, that the sum of residuals from $\boldsymbol{m}$ is identically null as seen in 5.12, we have:

$$\sum_{j=1}^{n}(x_j - m) = 0$$

$$\sum_{j=1}^{n}(x_j - \mu)^2 = \sum_{j=1}^{n}(x_j - m)^2 + n(m-\mu)^2$$

To go to variances, we have to take the expected values of bot members, remembering that the expected value of the sum is equal to the sum of expected values:

$$E\left[\sum_{j=1}^{n}(x_j - \mu)^2\right] = E\left[\sum_{j=1}^{n}(x_j - m)^2\right] + E\left[n(m-\mu)^2\right]$$

$$\sum_{j=1}^{n}E(x_j - \mu)^2 = E\left[\sum_{j=1}^{n}(x_j - m)^2\right] + nE(m-\mu)^2$$

Thereafter, when possible, we substitute to the expected values the variances s^2 and σ_m^2; and, for this last, the value given in 5.17:

$$n\sigma^2 = E\left[\sum_{j=1}^{n}(x_j - m)^2\right] + n\sigma_m^2$$

$$n\sigma^2 - n\frac{\sigma^2}{n} = \sigma^2(n-1) = E\left[\sum_{j=1}^{n}(x_j - m)^2\right]$$

Finally we can get the value of the variance obtained from experimental data:

$$\sigma^2 = E\left[\sum_{j=1}^{n}\frac{(x_j - m)^2}{n-1}\right]$$

Therefore $\sum_{j=1}^{n}\frac{(x_j - m)^2}{n-1}$ is a correct estimator of the variance[3] and we can put:

---

[3] While s^2 is a correct estimator of the variance σ^2, s is not a correct estimator of σ. We can have a correct estimator of σ using:

$$s_{corr} = \frac{\Gamma[(n-1)/2]}{\Gamma(n/2)}\sqrt{\frac{\sum_{i=1}^{n}(x_i - m_x)^2}{2}}$$

where $\Gamma(x) = \int_0^\infty e^{-\xi}\xi^{x-1}d\xi$.

$$s^2 = \sum_{j=1}^{n} \frac{(x_j - m)^2}{n-1} \tag{5.18}$$

5.4 TIPS ON ROBUST STATISTICS

The methods presented before have a very rigorous theoretical basis, that work perfectly in case of random variables. But in practice our experimental data contain always some results that are more or less affected by systematic effects and measurement accidents. To give an example of measurement accident, consider to have your measurement system on a table, working unattended automatically. Your colleague enters the room and shocks your system, without informing you about that. When you get the results you read 12, 11, 13, 12, 12, 25, 13, 12,12. Really strange to get a 25 in a group of data very near to 12! A more common example with the electronic high sensitivity instrument used nowadays, reports the effect of electromagnetic interference.

I was taking measurements with a very sensitive instrument (about 2 nV/V) in a workshop with machine tools and industrial arc welders in action. The instrument output was very stable. A telephone ringed in a room nearby. Suddenly the instrument became mad, the instrument went up and down hundreds of digits, until the telephone stopped ringing.

It was possible to understand the presence of the electromagnetic interference just because it was accompanied by the acoustical signal and produced an outstanding effect, but if it were silent and the instrument unattended, only the abrupt change of results would indicate what happened.

Generally speaking, the hypothesis of pure random variation, fundamental for the mathematical development given before, is never completely respected, therefore the results obtained as average and variance can be significantly wrong.

A real example can be seen in the results obtained on the speed of light made by Simon Newcomb in 1882. He measured the time required to a light signal to go from his laboratory to a mirror on the Washington Monument, and back.

The distance of the path travelled is 7.44373 km. In the first day of the experiment he gets the results reported in table 5.4.

Tab. 5.4 Newcomb's results, of the time in microseconds required to a light signal to fly a path of 7443.73 m. The first and the last results of the second column are strongly different from the others.

24.828	**24.756**	24.829	24.830
24.826	24.827	24.822	24.823
24.833	24.816	24.824	24.829
24.824	24.840	24.821	24.831
24.834	**24.798**	24.825	24.819

The average of these results is 24.822 μs and the standard deviation is 0.018 μs. But if the results strongly different from the others are considered measurement accidents and excluded, the average will change of 0.005 and the standard deviation reduced to 0.006. Very significant changes!

The presence of *outliers (i.e. outlying observations)*, results affected significantly by additional or different factors from factors affecting bulk of data, changes the results of the experiment. For this reason you can understand how useful can be to resort to other parameters less sensitive to outliers, that is to the field of so called *Robust Statistics*.

We will see the methods of robust statistics for calculating the functional equivalent of the average, called *Median*, and of the standard deviation, called *Inter Quartile Range* (*IQR*).

The first condition required is to put in order the data, as in table 5.5. Note that, formerly, it was a difficult operation, as there is not a calculation to do that and it was necessary to do it manually, sorting the results, but with modern computers, nowadays it is no more a problem.

Tab. 5.5 Newcomb's results in order. With 20 data, the median is a value between 10th and the 11th, 24.8255, first quartile is between 5th and 6th, 24.82175, and third quartile is between 15th and 16th, 24.82925. IQR is the difference between the first and third quartile, 0.0075.

24.756	24.822	24.826	24.830
24.798	24.823	24.827	24.831
24.816	24.824	24.828	24.833
24.819	24.824	24.829	24.834
24.821	24.825	24.829	24.840

The median is the value that in the ordered set lets an equal number of data before and after. In our case, with 20 data, we shall interpolate[4] between the 10th and the 11th data. Doing the interpolation we get the value 24.8255. If we do the same after excluding outliers, we get a median of 24.8265 with a change of 0.001, much smaller than the change 0.005 of the average.

The inter quartile range IQR can be evaluated considering quartiles. In general we can define quantiles, identified by a given percentage *C%*, that subdivide data range in two parts, one containing *C%* of the intervals between data and the other automatically containing 1- *C%* of the intervals. Quartiles subdivide data with *at* 25% intervals: the first quartile, *Q1*, has *C%*=25%, the second, *Q2*, is the median with *C%*=50%, the third, *Q3*, has *C%*=75%. IQR is the difference between the first and the third quartiles, *Q3-Q1*. In our case for

---

[4] Interpolation is done considering the number of intervals. In our case, 20 data mean 19 intervals. The median subdivide the intervals 50%-50%, therefore is posed at 19·50%=9.5 intervals, that is starting from data 1 at 10.5, the center between 10th and 11th data. For the first quartile the subdivision is 25%-75%, that is 19·25%=4.75, that is we have to interpolate 0.75 of the interval between 5th and 6th data.

the original data we have *Q1*=24.82175, *Q3*=24.82925, therefore we get a value of IQR=0.0075. After exclusion of the outliers we have IQR=0.0065, with a change of -15.4% referred to this last value considered more correct, while the corresponding change for standard deviation is -204%. It is evident how robust statistic can produce more reliable data. It is useful also for data presentation (descriptive statistics). A specific graphic presentation using median and quartiles is the box plot, as shown in Fig. 5.6a. On that base it is possible also to evidence empirically possible measurement accident. The classical way is to consider an outlier each data that is outside the interval from $Q1-1.5\cdot IQR$ to $Q3+1.5\cdot IQR$, as shown in Fig. 5.6b for Newcomb's results obtained in three different days.

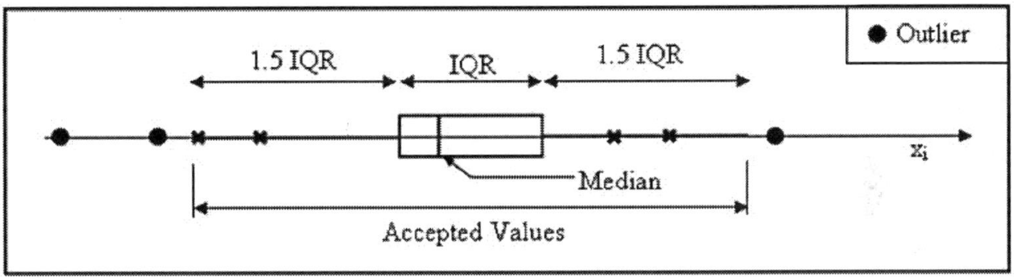

Fig. 5.6a Boxplot is a very useful graphic presentation based on robust statistics. The position index is given by the median, while dispersion index is taken from Q1 and Q3 quartiles, evidenced by a box. Data outside a conventional range of acceptable values are evidenced as outliers.

Medians give the position indexes of the three days, different, but the overlapping of the three boxes indicates that the three results are compatible. Two outliers are evidenced on the first day, but their effect on the median is not so strong so that the graphical presentation is useful and not misleading. About outliers detected with IQR, when the number of data is higher than 20, this method is very severe, if compared with other exclusion methods, therefore at Politecnico di Torino was proposed an empirical correction[5] with the values of limits in formula 5.19 that allows to have performances similar to other more complicated methods.

$$Q1-1.5\cdot \mathrm{IQR}\left[1+0.1\cdot\log\left(n/10\right)\right] \qquad Q3+1.5\cdot \mathrm{IQR}\left[1+0.1\cdot\log\left(n/10\right)\right] \qquad (5.19)$$

[5] G. Barbato, E. M. Barini, G. Genta, R. Levi (2011). "Features and performance of some outlier detection methods", Journal of Applied Statistics, vol. 38(10), pp. 2133-2149.

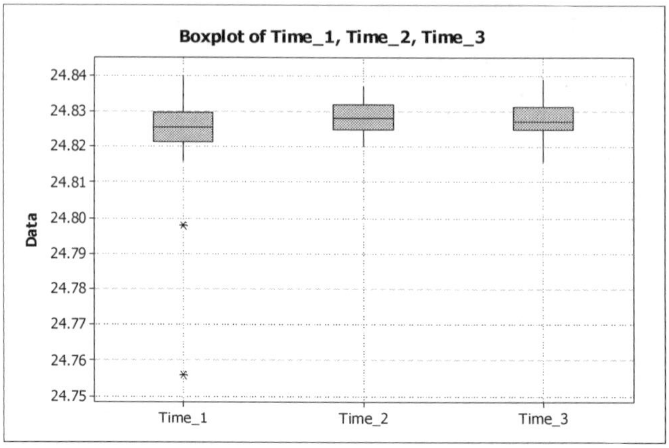

Fig. 5.6b The boxplot allows you to see the situation at a glance. Here Newcomb's results obtained in three different days are shown (data concerning the first day are those just analysed). One sees immediately that they are compatible. The use of medians allows a meaningful presentation even in presence of the two outliers highlighted on the first day.

5.5 THEORETICAL DISTRIBUTIONS

Some probability distributions describe common situations and for that reason they have been studied specifically. It is very efficient to have some theoretical bases, a good description of the hypotheses that produce a given distribution, to see if the real situation that we have to study can be represented adequately.

We will consider here only few distributions that are useful in verification activity and in data analysis including uncertainty evaluation.

5.5.1 Uniform distribution

Uniform distribution (also called rectangular distribution) was described with the example of a six-sided dice. It is widely applied in technical fields, representing, for example, what can happen within the reading resolution interval: your instruments can give the readings 1, 2, 3, ..., but when it read 1, the measurand can be anywhere between 0.5 and 1.5, as shown in Fig. 5.7, with no reason to be on one value more than any other value of that interval. Therefore, for the uniformity of the situation, the uniform distribution is applied.

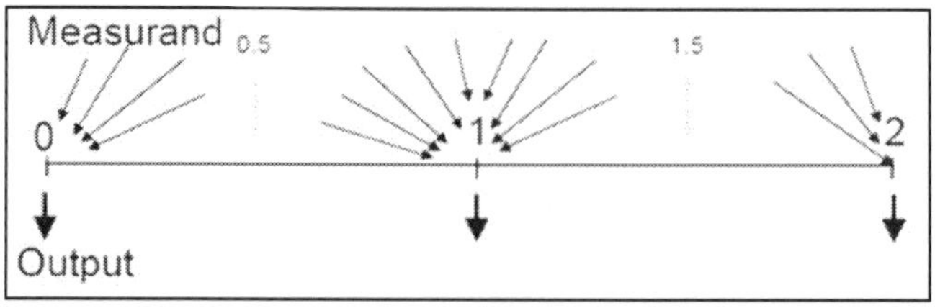

Fig. 5.7 Reading resolution effect. If your instrument output can be 0, 1, 2, ..., when you read 1, your measurand can be anywhere between 0.5 and 1.5..

GUM, for a matter of simplification, requires to use uniform distribution nearly in any case (but when the phenomenon involved is oscillating) when the real distribution in a bounded interval is unknown.

The uniform distribution is defined by its boundary, which can always be represented by symmetrical points $-a$ and a around zero (if necessary applying a coordinate system translation).

This defines the uniform density of probability: being a rectangle with the base equal to $2a$ and an area equal to 1, the height is $1/2a$.

The average, for symmetry reasons, is the central value, 0, and the variance can be easily calculated using equation 5.16:

$$\sigma^2 = \int_{-\infty}^{\infty} (x-\mu)^2 \cdot \mathrm{f}(x)dx = \int_{-a}^{a} \frac{x^2}{2a} dx = \left. \frac{x^3}{6a} \right|_{-a}^{a} = \frac{a^2}{3} \tag{5.20}$$

5.5.2 *Triangular distribution*

Also triangular distribution is quite common, being, as we have seen rolling two dice, the effect of two events with uniform distribution. We get the occasion to present a general method to evaluate the effect of two distributions, called convolution method.

5.5.2.1 Convolution method

Convolution method is really general, therefore to explain it we use a simple generic discrete distribution: $p(1)=1/6$, $p(2)=2/6$ and $p(3)=3/6$, as shown in Fig. 5.8. We can see what happen if we apply two times this distribution using the tree diagram given in Fig. 5.9. We obtain the calculus scheme represented in Fig. 5.10, which can be calculated with the formula 5.21:

$$P_s(x) = \sum p(y) \cdot p(x-y) \tag{5.21}$$

Being $x = 2, 3, 4, 5, 6$ the sum of outcomes and $y = 1, 2, 3$ the single outcome. It is easy to check that for x=5 we have (consider that $p(1) \cdot p(4)$ is zero as $p(4)=0$):

$$P_s(5) = p(2) \cdot p(3) + p(3) \cdot p(2)$$

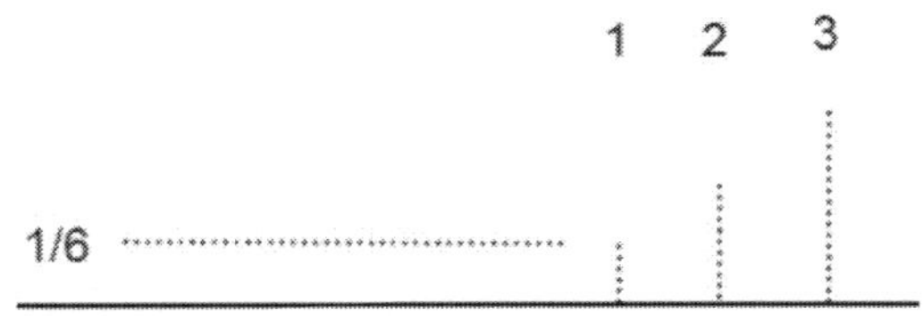

Fig. 5.8 Simple discrete distribution $p(1)=1/6$, $p(2)=2/6$ and $p(3)=3/6$, used for explaining convolution.

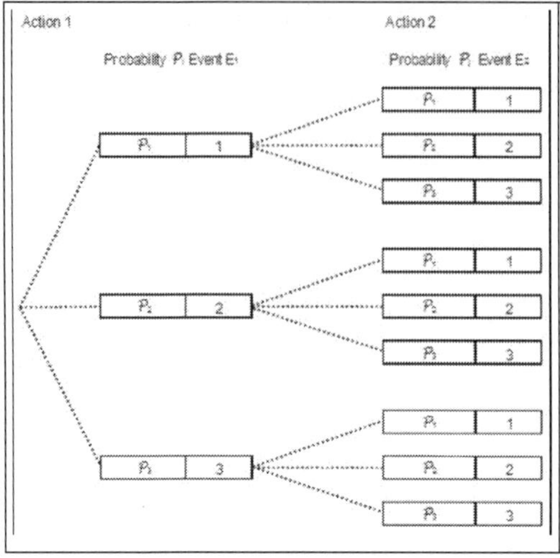

Probability E_1+E_2	$\mathcal{P}_x$		
2	$\mathcal{P}_1 * \mathcal{P}_1$		
3	$\mathcal{P}_1 * \mathcal{P}_2$	+ $\mathcal{P}_2 * \mathcal{P}_1$	
4	$\mathcal{P}_1 * \mathcal{P}_3$	+ $\mathcal{P}_2 * \mathcal{P}_2$	+ $\mathcal{P}_3 * \mathcal{P}_1$
5	$\mathcal{P}_2 * \mathcal{P}_3$	+ $\mathcal{P}_3 * \mathcal{P}_2$	
6	$\mathcal{P}_3 * \mathcal{P}_3$		

Fig. 5.9
Tree diagram for two time application of the given distribution.

Fig. 5.10
Calculation scheme for the probability of the sum.

This type of calculus can be described as putting on two parallel coordinate axes the given distribution on the first axis and the second distribution in the opposite direction as shown in Fig. 5.11.

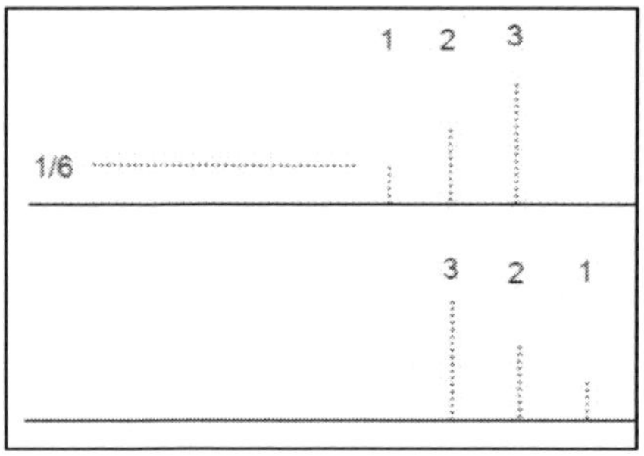

Fig. 5.11 Scheme for calculating the probability of the sum. In this position the sum of outcomes of the upper part and the lower part is 5 and the corresponding sum of products of coincident positions give the result $\mathcal{P}_s(5)$ given in Fig. 5.10.

In the position showed the sum of outcomes of the upper part and the lower part is 5 and the corresponding sum of products of coincident positions give the result $\mathcal{P}_s(5)$ given in Fig. 5.10.

If you shift the lower axis one step on the left, you get the sum of outcomes equal to 4, and doing the same calculations as before you get the value of $\mathcal{P}_s(4)$. Therefore a method for calculating the distribution of the sum of outcomes consists to shift the lower axis in each possible position.

This is valid also for continuous distributions, if you change summation in integral. The formula, which is a particular case of composition products considered by the Italian mathematician Vito Volterra in 1913, is given here:

$$r(x) = \int_{-\infty}^{\infty} f(y) \cdot g(x-y) \, dy \tag{5.22}$$

where $f(y)$ is the fixed distribution, $g(x-y)$ is the moving distribution while $r(x)$ is the convolution distribution.
If we apply this to the uniform distribution bounded at $-a$ and a as shown in Fig. 5.12 we shall consider that for $x<0$ r(x) is increasing while for $x>0$ is decreasing, therefore we divide the integration in two parts.

We shall also evaluate carefully the values of y as a function of x that bounds the interval where $f(x)\cdot g(x-y)$ is not null.

Performing the calculations given hereafter:

$$-2a<x<0$$
$$f(y)\cdot g(x-y)\neq 0 \qquad -a<y<x+a$$
$$r(x)=\int_{-\infty}^{\infty} f(y)\cdot g(x-y)\,dy=\int_{-a}^{x+a}\frac{1}{2a}\frac{1}{2a}\,dy$$
$$r(x)=\left.\frac{y}{4a^2}\right|_{-a}^{x+a}=\frac{x+a+a}{4a^2}=\frac{x+2a}{4a^2}$$

$$0<x<2a$$
$$f(y)\cdot g(x-y)\neq 0 \qquad x-a<y<a$$
$$r(x)=\left.\frac{y}{4a^2}\right|_{x-a}^{a}=\frac{a-x+a}{4a^2}=\frac{2a-x}{4a^2}$$

we get the solution of equations 5.23:

$$-2a<x<0 \qquad r(x)=\frac{x+2a}{4a^2}$$
$$0<x<2a \qquad r(x)=\frac{2a-x}{4a^2} \tag{5.23}$$

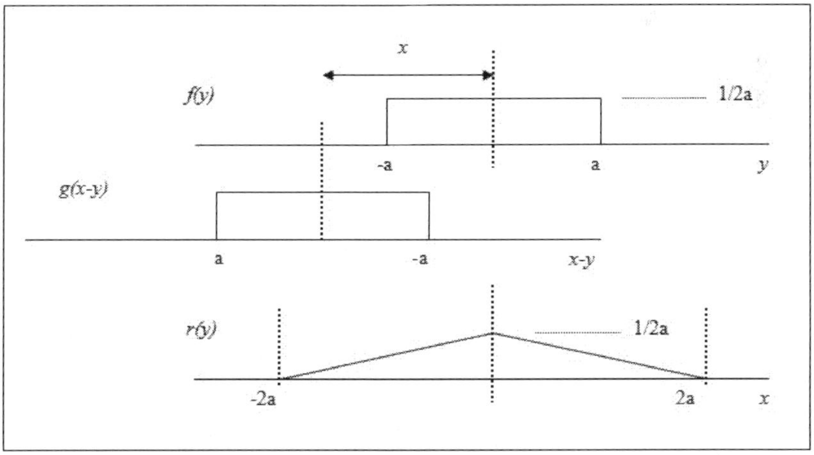

Fig. 5.12 Convolution of two uniform distributions, producing a triangular distribution.

GUM considers the triangular distribution bounded at $-a$ and a and, accordingly, we can transform immediately the equation found before.

$$-a < x < 0 \qquad \mathrm{r}(x) = \frac{x+a}{a^2} \qquad\qquad 0 < x < a \qquad \mathrm{r}(x) = \frac{a-x}{a^2}$$

$$\sigma^2 = \int_{-\infty}^{\infty} (x-\mu)^2 \cdot \mathrm{r}(x) \cdot \mathrm{d}x = \int_{-a}^{0} x^2 \cdot \frac{x+a}{a^2} \cdot \mathrm{d}x + \int_{0}^{a} x^2 \cdot \frac{a-x}{a^2} \cdot \mathrm{d}x$$

$$\sigma^2 = \frac{1}{a^2}\left[\int_{-a}^{0} \left(x^3 + ax^2\right)\mathrm{d}x + \int_{0}^{a} \left(ax^2 - x^3\right)\mathrm{d}x\right] = \frac{1}{a^2}\left[\left(\frac{x^4}{4} + a\frac{x^3}{3}\right)\Bigg|_{-a}^{0} + \left(a\frac{x^3}{3} - \frac{x^4}{4}\right)\Bigg|_{0}^{a}\right]$$

$$\sigma^2 = \frac{1}{a^2}\left[-\left(\frac{a^4}{4} - a\frac{a^3}{3}\right)\Bigg|_{-a}^{0} + \left(a\frac{a^3}{3} - \frac{a^4}{4}\right)\Bigg|_{0}^{a}\right] = 2a^2\left(\frac{1}{3} - \frac{1}{4}\right) = 2a^2\frac{4-3}{12} = \frac{a^2}{6}$$

In this condition the average is μ=0 and the variance, as calculated in the same equation 5.24, is $a^2/6$.

$$\sigma^2 = \frac{a^2}{6} \tag{5.24}$$

5.5.3 Arcsine distribution

The arcsine distribution is used by GUM, under the name of *U* distribution, because it describes what happen with oscillating phenomena. Natural oscillating phenomena, like vibrations, alternate mechanical movements, effects of rotations, etc., have a typical characteristic to be in positions near the minimum or the maximum a longer time than in the central position. This involves that if you do an action at random time, it is more likely to get an oscillating phenomenon in its maximum or minimum condition. The probability density function has, therefore a *U* shape, going from the minimum condition, where it is high, to the centre, where it is low, to the maximum condition, where it is high again. To theorize on oscillating phenomena a sinusoidal oscillation is used: the outcome θ is a sinusoidal function of the time t:

$$\theta = a \cdot \sin 2\pi \frac{t}{T}$$

We can consider half a period, $T/2$, of that function, as shown in Fig. 5.13, and being uniform the probability to take your action throughout that time interval, you evaluate the probability $\mathrm{f}(t)\mathrm{d}t$ to do your action in any time interval $\mathrm{d}t$ of $T/2$ that is clearly equal, for the monotonic trend of the function, to the probability of getting an outcome in the interval $\mathrm{d}\theta$:

$$\mathrm{f}(t)\,\mathrm{d}t = \frac{2\,\mathrm{d}t}{T} = g(\theta)\,\mathrm{d}\theta$$

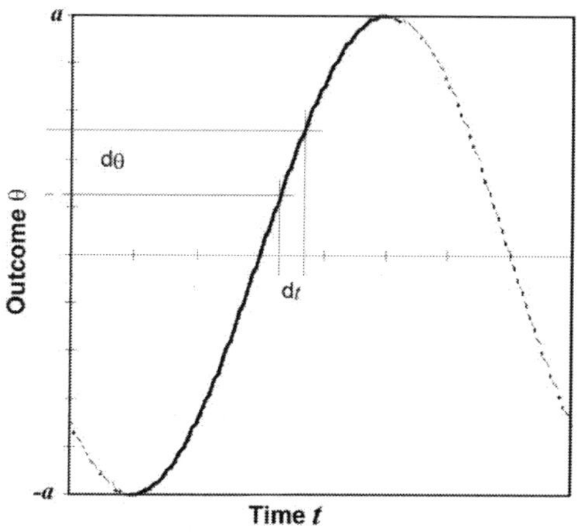

Fig. 5.13 Oscillating phenomenon represented by a sinusoidal function.

Thereafter we can evaluate the density of probability for the variable θ:

$$g(\theta) = \frac{2}{T}\frac{\mathrm{d}t}{\mathrm{d}\theta}$$

but being:

$$t = \frac{T}{2\pi} arcsen \frac{\theta}{a}$$

$$\frac{\mathrm{d}t}{\mathrm{d}\theta} = \frac{T}{2\pi\sqrt{a^2 - \theta^2}}$$

we obtain the formula of the density of probability for the arcsine distribution in equation 5.25:

$$g(\theta) = \frac{1}{\pi\sqrt{a^2 - \theta^2}} \tag{5.25}$$

Also in this case the average is zero, for symmetry, and the variance is calculated using equation 5.16:

$$\sigma^2 = \int_{-a}^{a} \frac{\theta^2}{\pi\sqrt{a^2 - \theta^2}} \cdot \mathrm{d}\theta = \frac{1}{\pi}\left[\frac{a^2}{2}\arcsin\left(\frac{\theta}{a}\right) - \frac{\theta\sqrt{a^2 - \theta^2}}{2}\right]\Bigg|_{-a}^{a} =$$

$$= \frac{1}{\pi}\left[\frac{a^2}{2}\arcsin\left(\frac{a}{a}\right) - \frac{a\sqrt{a^2 - a^2}}{2}\right] - \frac{1}{\pi}\left[\frac{a^2}{2}\arcsin\left(\frac{-a}{a}\right) + \frac{a\sqrt{a^2 - a^2}}{2}\right] = \frac{a^2}{2\pi}\left[\frac{\pi}{2} + \frac{\pi}{2}\right]$$

and the result is given in equation 5.26

$$\sigma^2 = \frac{a^2}{2} \tag{5.26}$$

5.5.4 Binomial distribution

A very common procedure in verification, for instance, of production batches consists in taking number n of samples from the batch and checking if they are in tolerance or out of tolerance. The acceptance conditions, usually establish the number of pieces out of tolerance k that can be accepted. Generally, in industrial testing, when a sample is tested is not put back in the batch, therefore we are in the condition of a card deck; if you draw an ace, it is no more available in a second drawing. Therefore the probability of getting an ace of hearths at a very first drawing from a deck of 52 cards is 4/52, if you get it, at the second drawing the probability has changed to 3/51. A change of probabilities in the successive drawings characterizes this working condition that is described as drawing without replacement. This condition is more complicated than the condition of the outcomes of a dice: if you get 5 at the first roll, you have the same probability of getting 5 at the subsequent rolls. Let us start with this less complicate condition that can be exemplified by a test of a measuring instrument: if you get a result out of tolerance this will not change the probability of getting a result out of tolerance in subsequent trials. This is called *drawing with replacement*, and in this case the probability of the selected event does not change, let us call it p the probability of getting the outcome out of tolerance (OT) in a single trial. Consider we repeat the trial three times, that is n=3 and we want to see the probability of getting with all the trials 0, 1, 2 or 3 outcomes out of tolerance, that is we want the statistical distribution against the variable k, number of out of tolerance. We can solve this problem using the tree diagram, as shown in Fig. 5.14.

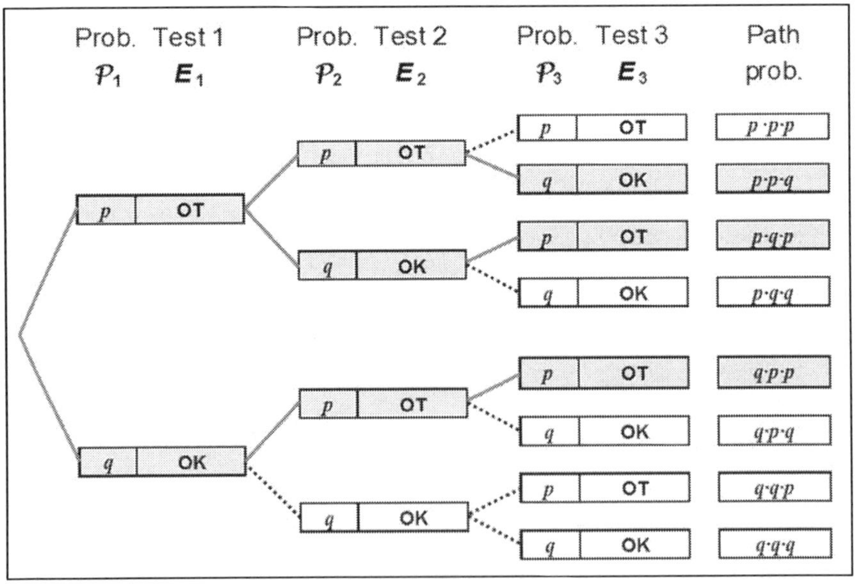

Fig. 5.14 Tree diagram for n=3 trials with probability p of getting a piece out of tolerance (OT) and q=1-p to get a piece in tolerance (OK). In yellow the paths for k=2.

The situation k=2 can be attained following three paths. Why three paths? Because the number of combinations of the 3 trials for getting 2 OT is 3. You can check that if you require only one OT you have 3 paths as the combinations of three, one at a time, while if you want to have 0 OT or 3 OT you have only one path as the relevant combinations. Why each path for 2 OT has a path probability $p \cdot p \cdot q$? Because you ask to have exactly 2 events with probability p, therefore the last event with probability q and you have the product of corresponding probabilities. Check, for instance, that if you ask for 1 OT you have path probability $p \cdot q \cdot q$.

From what we have seen we can find a general formula (5.27):

$$\mathcal{P}(n,k) = \binom{n}{k} p^k q^{n-k} = \frac{n!}{k!(n-k)!} p^k q^{n-k} \tag{5.27}$$

It is useful to know the position index and dispersion index of our distribution. We start with the special case of a single trial, that is n=1 and k equal to 0 or 1. That distribution is called Bernoulli Distribution.

We use the general equations 5.4 and 5.14:

$$\mu = \sum_{k=0}^{1} k\mathcal{P}(k) = \sum_{k=0}^{1} kp^k q^{n-k} = 0 \cdot q + 1 \cdot p = p \tag{5.28}$$

And for the variance:

$$\begin{aligned} \sigma^2 &= \sum_{k=0}^{1} (k-\mu)^2 \mathcal{P}(k) = \sum_{k=0}^{1} (k-\mu)^2 p^k q^{n-k} = (0-p)^2 \cdot q + (1-p)^2 \cdot p = \\ &= p^2 q + q^2 p = pq(p+q) = pq \end{aligned} \tag{5.29}$$

If more than one trial is made, can be considered as a one trial repeated, and being the average and the variance additive, for n (independent) trials we have equations 5.30:

$$\mu = np \qquad\qquad \sigma^2 = npq \tag{5.30}$$

Examples of binomial distributions with n=20 and different values of p are given in Fig. 5.15.

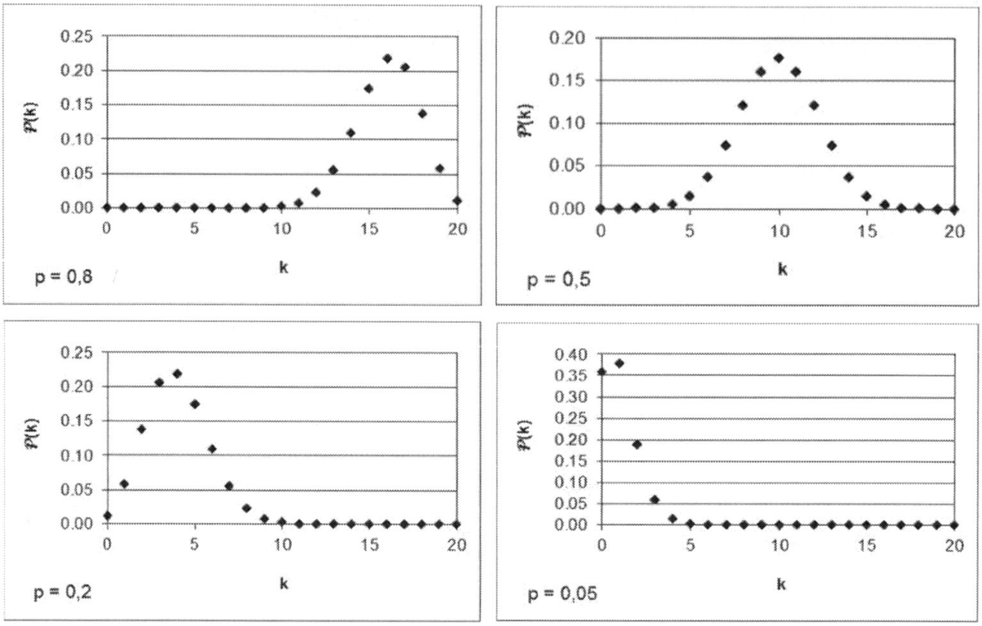

Fig. 5.15 Graphs of binomial distributions obtained for n=20 and some values of p.

5.5.5 Hypergeometric distribution

As said before, the binomial distribution represents well the condition of drawing with replacement, but frequently the industrial testing condition is represented by the condition of *drawing without replacement,* therefore binomial distribution can be used only as an approximation when the events probability do not change significantly in successive tests. To represent correctly the condition of drawing without replacement the hypergeometric distribution shall be used.

Let us describe what happen testing a batch of workpieces, if M is the number of workpieces in the batch, C the number of workpieces in a selected condition (usually workpieces out of tolerance, OT), n the number of samples taken for the test and k the number of workpieces in the selected condition within the n samples. At the first drawing, as described by the tree diagram in Fig. 5.16, we have a probability C/M to get a piece OT (in the selected condition), but at the second drawing the number of pieces in the batch has changed to $M-1$, and if you have got a piece OT, their number is changed to $C-1$, as you can see from the tree diagram. At each subsequent step the number of pieces decrease of one unit, and when you take a piece OT also their number decrease of one unit. We stop our tree diagram at the third drawing looking for a general rule valid for any value of n and k. From the tree diagram it is easy to evaluate the probabilities to find 0, 1, 2 or 3 pieces OT.

$$\mathcal{P}(0)=\frac{(M-C)(M-C-1)(M-C-2)}{M(M-1)(M-2)} \qquad \mathcal{P}(1)=3\frac{(M-C)(M-C-1)C}{M(M-1)(M-2)}$$

$$\mathcal{P}(2) = 3\frac{(M-C)C(C-1)}{M(M-1)(M-2)} \qquad \mathcal{P}(3) = \frac{C(C-1)(C-2)}{M(M-1)(M-2)}$$

These formulas can be generalized with the same method used for the binomial distribution: the number of paths for each value of k corresponds to the number of combinations of n at k at a time, and each path for given n and k has a constant probability:

$$\mathcal{P}(k) = \binom{n}{k}\frac{C(C-1)\ldots(C-k+1)(M-C)(M-C-1)\ldots(M-C-n+k+1)}{M(M-1)\ldots(M-n+1)}$$

Thereafter, with some mathematical management it is possible to get the compact formula given in equation 5.31:

$$\mathcal{P}(k) = \frac{\binom{C}{k}\binom{M-C}{n-k}}{\binom{M}{n}} \tag{5.31}$$

It can be useful to have the values of average and variance:

$$\mu = n\frac{C}{M} \qquad \sigma^2 = n\frac{C(M-C)(M-n)}{M^2(M-1)} \tag{5.32}$$

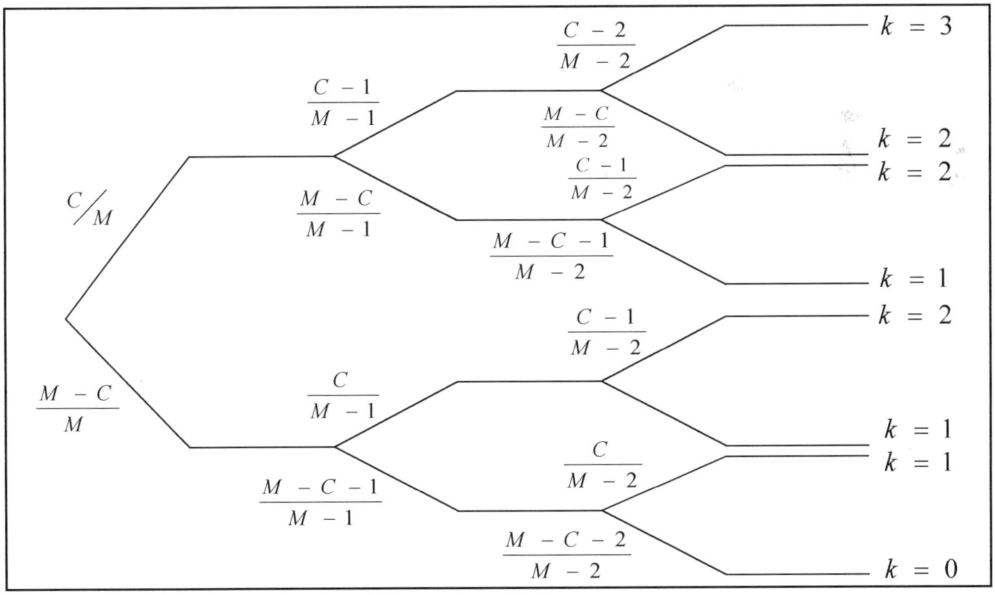

*Fig. 5.16 Tree diagram for a testing process corresponding to drawing without replacement for a batch of **M** pieces and **C** pieces in a selected condition. Number of drawings **n**=3. To avoid filling the tree diagram with writing of event conditions, we used the convention that the branches going up correspond to happening of the selected event, while the branch going down to the contrary.*

Binomial and hypergeometric distributions are useful in many practical situations; however, the calculation of factorials, although trivial, is, in practice, very heavy. To reduce the computational problems, two important approximations of the binomial distribution were developed, both valid for n large. We will discuss before Poisson distribution, which is a valid approximation for very low values of p (therefore sometimes called law of rare events), thereafter the normal distribution (often called De Moivre-Laplace or Gauss distribution), valid for all values of p, although provides better results for central values ($p \approx 0.5$).

5.5.6 Poisson distribution

We examine, now, the binomial distribution in the case of very small p (distribution of rare events) and n very large. An important parameter is the product $np = \lambda$. In this case the binomial distribution can be represented by the approximate formula:

$$\mathcal{P}(n,k)=\binom{n}{k}p^k q^{n-k}=\frac{n!}{k!(n-k)!}p^k q^{n-k}=\frac{\lambda^k e^{-\lambda}}{k!} \tag{5.33}$$

This approximation, called Poisson distribution, Fig. 5.17, has the advantage to avoid the calculation of $n!$ that is very complicated.

The mean value and the variance of Poisson distribution can be approximated from the values of the binomial distribution:

$$\mu = np = \lambda \qquad \sigma_r^2 = npq = np(1-p) \approx np = \lambda \tag{5.34}$$

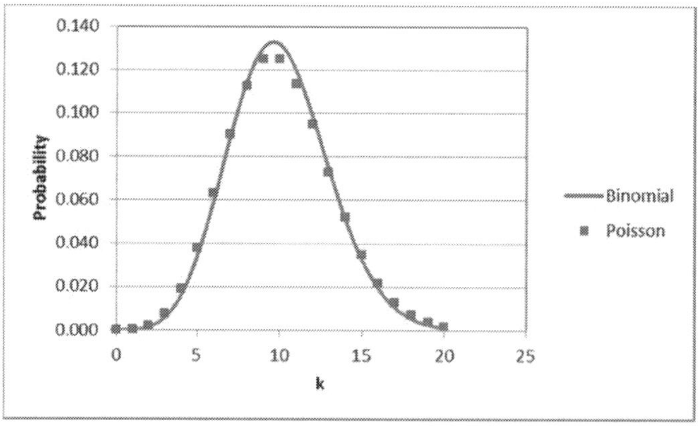

Fig. 5.17 Graphs of Poisson distribution compared with binomial distribution obtained for n=100 and p=0.1. The approximation is questionable because n is not sufficiently high and p not sufficiently small.

5.5.7 Normal distribution

In the field of experimental applications, the normal distribution is crucial for its characteristic to describe what happens as normal effect of chance and this has earned it, in fact, its name.

In the case of the experimental data, the basic hypothesis is linked to the fact that great care is always taken to avoid systematic errors and measurement accidents, therefore the main effect is due to small comparable effects of random factors. This case is considered by the Central Limit Theorem (CLT). Avoiding formal mathematical description we can report the "fuzzy" central limit theorem which says that data influenced by many small and independent random effects are approximately normally distributed. Let us examine what can happen by combining a certain number of random factors. For example, we have seen what happens taking the sum of outcomes of two dice (Fig. 5.1). We can see now what happen with three dice, as shown in Fig. 5.18: one can clearly see the bell shape.

								6-3-1	6-4-1							
								6-2-2	6-3-2							
							6-2-1	6-1-3	6-2-3	6-5-1						
							6-1-2	5-4-1	6-1-4	6-4-2						
							5-3-1	5-3-2	5-5-1	6-3-3						
							5-2-2	5-2-3	5-4-2	6-2-4						
						6-1-1	5-1-3	5-1-4	5-3-3	6-1-5	6-6-1					
						5-2-1	4-4-1	4-5-1	5-2-4	5-6-1	6-5-2					
						5-1-2	4-3-2	4-4-2	5-1-5	5-5-2	6-4-3					
						4-3-1	4-2-3	4-3-3	4-6-1	5-4-3	6-3-4					
						4-2-2	4-1-4	4-2-4	4-5-2	5-3-4	6-2-5					
						4-1-3	3-5-1	4-1-5	4-4-3	5-2-5	6-1-6					
					5-1-1	3-4-1	3-4-2	3-6-1	4-3-4	5-1-6	5-6-2	6-6-2				
					4-2-1	3-3-2	3-3-3	3-5-2	4-2-5	4-6-2	5-5-3	6-5-3				
					4-1-2	3-2-3	3-2-4	3-4-3	4-1-6	4-5-3	5-4-4	6-4-4				
					3-3-1	3-1-4	3-1-5	3-3-4	3-6-2	4-4-4	5-3-5	6-5-3				
					3-2-2	2-5-1	2-6-1	3-2-5	3-5-3	4-3-5	5-2-6	6-6-2				
				4-1-1	3-1-3	2-4-2	2-5-2	3-1-6	3-4-4	4-2-6	4-6-3	5-6-3	6-6-3			
Outcomes				3-2-1	2-4-1	2-3-3	2-4-3	2-6-2	3-3-5	3-6-3	4-5-4	5-5-4	6-5-4			
				3-1-2	2-3-2	2-2-4	2-3-4	2-5-3	3-2-6	3-5-4	4-4-5	5-4-5	6-4-5			
				2-3-1	2-2-3	2-1-5	2-2-5	2-4-4	2-6-3	3-4-5	4-3-6	5-3-6	6-3-6			
			3-1-1	2-2-2	2-1-4	1-6-1	2-1-6	2-3-5	2-5-4	3-3-6	3-6-4	4-6-4	5-6-4	6-6-4		
			2-2-1	2-1-3	1-5-1	1-5-2	1-6-2	2-2-6	2-4-5	2-6-4	3-5-5	4-5-5	5-5-5	6-5-5		
			2-1-2	1-4-1	1-4-2	1-4-3	1-5-3	1-6-3	2-3-6	2-5-5	3-4-6	4-4-6	5-4-6	6-4-6		
		2-1-1	1-3-1	1-3-2	1-3-3	1-3-4	1-4-4	1-5-4	1-6-4	2-4-6	2-6-5	3-6-5	4-6-5	5-6-5	6-6-5	
		1-2-1	1-2-2	1-2-3	1-2-4	1-2-5	1-3-5	1-4-5	1-5-5	1-6-5	2-5-6	3-5-6	4-5-6	5-5-6	6-5-6	
	1-1-1	1-1-2	1-1-3	1-1-4	1-1-5	1-1-6	1-2-6	1-3-6	1-4-6	1-5-6	1-6-6	2-6-6	3-6-6	4-6-6	5-6-6	6-6-6
Sum	**3**	**4**	**5**	**6**	**7**	**8**	**9**	**10**	**11**	**12**	**13**	**14**	**15**	**16**	**17**	**18**

Fig. 5.18 Effect of combining three factors independent and with equal variability: it is evident the bell shape similar to a normal distribution.

Practically, combining three independent factors, even if they have a quite unnatural distribution like the uniform distribution of one dice outcome, you get a result that is not very far from a normal distribution. Therefore, for many practical situations, we can say that the composition of at least three independent factors produces a statistical distribution that can be managed as a normal distribution. Some attention is necessary: if on the one side the distribution shape of the factors is not very important, what is fundamental is that they are really three or more independent factors. Therefore, if you have three, or also 10 factors among which one has a large variability and all the others a small variability, you will see only the effect of the large variability factor, therefore you get the effect of a single factor, and your composition will be very similar to that of the single large variability factor, not significantly modified by the effects of other factors. The same happens if your factors are connected one to the other (namely *correlated*, that is a condition not so strong as the independence, but such that it excludes independence) within their variability interval[6], that is if when one randomly increases, also the others increase or when one decreases, also the others decrease. These two conditions shall be considered carefully but when are absent, that is, when at least three non-correlated factors of compatible variability are combined, the results are normally distributed.

A method to derive the equation for the probability density distribution is to find the normal approximation to the binomial distribution, discrete, for a number of trials n tending to infinity and p=0.5. Equation 5.35 is thus obtained for the density of probability of normal distribution:

$$f(x)=\frac{1}{\sigma\sqrt{2\pi}}e^{-\frac{1}{2}\left(\frac{x-\mu}{\sigma}\right)^2} \tag{5.35}$$

where μ is the average and σ the standard deviation.

Now we shall note that the normal distribution is not discrete but continuous (as the unit is very small compared to ∞) and that the translation to the mean position given by x-μ, being μ=∞/2 corresponds to change the field of existence of the binomial (0<k<∞) to the field of existence of the normal -∞<x<∞. We note, as shown in Fig. 5.19, that the characteristic bell shape is positioned around the value of μ and is more or less large depending on the value of σ.

For managing the density of probability of normal distribution and the corresponding cumulated probability given in equation 5.36:

$$F(x)=\int_{-\infty}^{x} f(\xi)\,d\xi \tag{5.36}$$

[6] Frequently we manage factors that have a physical connection; it seems natural in that case to consider the factors involved as correlated. That is usually not true. One shall compare the variations due to the physical relationship with random variations of other factors involved. It is important to underline that we are interested in variations that happen in the variability interval of the phenomenon we are observing (e.g. variability intervals are the uncertainties in case of measurements, tolerances in case of production). Variability intervals are generally very small, such that the physical connection is frequently negligible when compared with other independent random effects. Nevertheless you shall be careful when the same random effect affects two variables, because in that case you can significant random effects correlated between the two variables. A common case is when two quantities involved in the observed phenomenon are measured with the same instrument; in that case you must check that the random effect produced on both variables by the measuring instrument do not involve a significant correlation.

it was necessary to use numerical tables, e.g. table 5.6 that refers to a so called standardized variable of the normal distribution z that using a translation $z - \mu$, and a variation of the scale by the factor σ transforms any distribution in the standardized normal distribution having μ=0 and σ=1:

$$z = \frac{x - \mu}{\sigma} \tag{5.37}$$

so that:

$$f(z) = \frac{1}{\sqrt{2\pi}} e^{-\frac{z^2}{2}} \tag{5.38}$$

$$F(z) = \int_{-\infty}^{z} f(\varsigma)\, d\varsigma \tag{5.39}$$

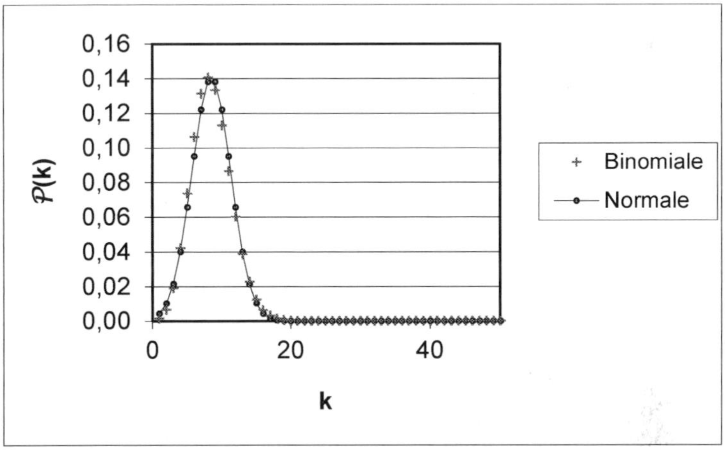

Fig. 5.19 Comparison between binomial and normal distribution for n *= 170 e* $\mathcal{P}$ *= 0.05. Even if* $\mathcal{P}$ *is really different from the nominal 0.5, the agreement is good and shows how it is possible to pass from a discrete distribution with a positive field of existence to a continuous distribution defined from -∞ to +∞*

The passage from any normal distribution N(μ, σ) to the standardized one is simple: one calculates z and enters in table 5.6 for getting F(z). The passage from f(z) and F(z) to f(x) and F(x) can be easily done considering the transformation of the scales, as seen in equations 5.40:

$$\mathrm{f}(x) = \frac{\mathrm{f}(z)}{\sigma} \quad \text{e} \quad \mathrm{F}(x) = \mathrm{F}(z) \tag{5.40}$$

Cumulative probability distribution function, e.g. F(z), gives the probability that the values of the variable of the distribution, e.g. σ, be between -∞ and a selected value, e,g, z.

The same is valid for any statistical distribution, normal standardized, simple normal, but also uniform, binomial, etc.

For continuous distributions the integral given in equation 5.39 can be represented by the area under the function of probability distribution between $-\infty$ and a selected value, as represted in Fig. 5.20.

Tab. 5.6 Values of the normal distribution cumulated probability $F(z)$ at the intersection of the line containing the integer part and the first decimal place of z, and the column containing the second decimal place. For negative values of z, $F(-z) = 1 - F(z)$.

Z	0.00	0.01	0.02	0.03	0.04	0.05	0.06	0.07	0.08	0.09
0.0	0.5000	0.5040	0.5080	0.5120	0.5160	0.5199	0.5239	0.5279	0.5319	0.5359
0.1	0.5398	0.5438	0.5478	0.5517	0.5557	0.5596	0.5636	0.5675	0.5714	0.5753
0.2	0.5793	0.5832	0.5871	0.5910	0.5948	0.5987	0.6026	0.6064	0.6103	0.6141
0.3	0.6179	0.6217	0.6255	0.6293	0.6331	0.6368	0.6406	0.6443	0.6480	0.6517
0.4	0.6554	0.6591	0.6628	0.6664	0.6700	0.6736	0.6772	0.6808	0.6844	0.6879
0.5	0.6915	0.6950	0.6985	0.7019	0.7054	0.7088	0.7123	0.7157	0.7190	0.7224
0.6	0.7257	0.7291	0.7324	0.7357	0.7389	0.7422	0.7454	0.7486	0.7517	0.7549
0.7	0.7580	0.7611	0.7642	0.7673	0.7704	0.7734	0.7764	0.7794	0.7823	0.7852
0.8	0.7881	0.7910	0.7939	0.7967	0.7995	0.8023	0.8051	0.8078	0.8106	0.8133
0.9	0.8159	0.8186	0.8212	0.8238	0.8264	0.8289	0.8315	0.8340	0.8365	0.8389
1.0	0.8413	0.8438	0.8461	0.8485	0.8508	0.8531	0.8554	0.8577	0.8599	0.8621
1.1	0.8643	0.8665	0.8686	0.8708	0.8729	0.8749	0.8770	0.8790	0.8810	0.8830
1.2	0.8849	0.8869	0.8888	0.8907	0.8925	0.8944	0.8962	0.8980	0.8997	0.9015
1.3	0.9032	0.9049	0.9066	0.9082	0.9099	0.9115	0.9131	0.9147	0.9162	0.9177
1.4	0.9192	0.9207	0.9222	0.9236	0.9251	0.9265	0.9279	0.9292	0.9306	0.9319
1.5	0.9332	0.9345	0.9357	0.9370	0.9382	0.9394	0.9406	0.9418	0.9429	0.9441
1.6	0.9452	0.9463	0.9474	0.9484	0.9495	0.9505	0.9515	0.9525	0.9535	0.9545
1.7	0.9554	0.9564	0.9573	0.9582	0.9591	0.9599	0.9608	0.9616	0.9625	0.9633
1.8	0.9641	0.9649	0.9656	0.9664	0.9671	0.9678	0.9686	0.9693	0.9699	0.9706
1.9	0.9713	0.9719	0.9726	0.9732	0.9738	0.9744	0.9750	0.9756	0.9761	0.9767
2.0	0.9772	0.9778	0.9783	0.9788	0.9793	0.9798	0.9803	0.9808	0.9812	0.9817
2.1	0.9821	0.9826	0.9830	0.9834	0.9838	0.9842	0.9846	0.9850	0.9854	0.9857
2.2	0.9861	0.9864	0.9868	0.9871	0.9875	0.9878	0.9881	0.9884	0.9887	0.9890
2.3	0.9893	0.9896	0.9898	0.9901	0.9904	0.9906	0.9909	0.9911	0.9913	0.9916
2.4	0.9918	0.9920	0.9922	0.9925	0.9927	0.9929	0.9931	0.9932	0.9934	0.9936
2.5	0.9938	0.9940	0.9941	0.9943	0.9945	0.9946	0.9948	0.9949	0.9951	0.9952
2.6	0.9953	0.9955	0.9956	0.9957	0.9959	0.9960	0.9961	0.9962	0.9963	0.9964
2.7	0.9965	0.9966	0.9967	0.9968	0.9969	0.9970	0.9971	0.9972	0.9973	0.9974
2.8	0.9974	0.9975	0.9976	0.9977	0.9977	0.9978	0.9979	0.9979	0.9980	0.9981
2.9	0.9981	0.9982	0.9982	0.9983	0.9984	0.9984	0.9985	0.9985	0.9986	0.9986
3.0	0.9987	0.9987	0.9987	0.9988	0.9988	0.9989	0.9989	0.9989	0.9990	0.9990
3.1	0.9990	0.9991	0.9991	0.9991	0.9992	0.9992	0.9992	0.9992	0.9993	0.9993
3.2	0.9993	0.9993	0.9994	0.9994	0.9994	0.9994	0.9994	0.9995	0.9995	0.9995
3.3	0.9995	0.9995	0.9995	0.9996	0.9996	0.9996	0.9996	0.9996	0.9996	0.9997
3.4	0.9997	0.9997	0.9997	0.9997	0.9997	0.9997	0.9997	0.9997	0.9997	0.9998
3.5	0.9998	0.9998	0.9998	0.9998	0.9998	0.9998	0.9998	0.9998	0.9998	0.9998
3.6	0.9998	0.9998	0.9999	0.9999	0.9999	0.9999	0.9999	0.9999	0.9999	0.9999
3.7	0.9999	0.9999	0.9999	0.9999	0.9999	0.9999	0.9999	0.9999	0.9999	0.9999
3.8	0.9999	0.9999	0.9999	0.9999	0.9999	0.9999	0.9999	0.9999	0.9999	0.9999
3.9	1.0000	1.0000	1.0000	1.0000	1.0000	1.0000	1.0000	1.0000	1.0000	1.0000
4.0	1.0000	1.0000	1.0000	1.0000	1.0000	1.0000	1.0000	1.0000	1.0000	1.0000

It is also simple to calculate the probability that the variable value is in a given interval, e.g. that z is between a e b, from the difference **F(b) - F(a)**:

$$\mathcal{P}(a<z<b) = F(b) - F(a) \qquad (5.41)$$

Even if nowadays Statistical tables are nearly not used, as the main statistical functions are available on any spreadsheet, we give here two examples on the main use of the table 5.6 to better understand the properties of normal distribution; to evaluate the probability related to a given interval of the variable, or, given a level of probability, to determine the boundary values of the corresponding variable interval.

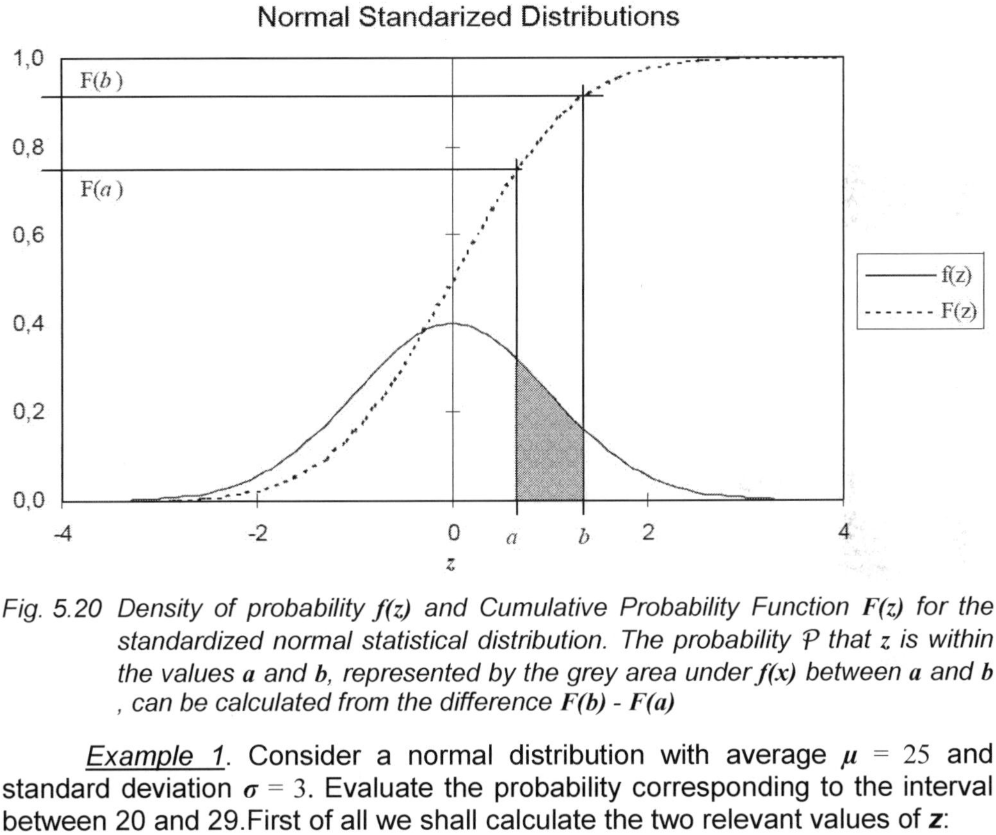

Fig. 5.20 Density of probability $f(z)$ and Cumulative Probability Function $F(z)$ for the standardized normal statistical distribution. The probability $\mathcal{P}$ that z is within the values a and b, represented by the grey area under $f(x)$ between a and b , can be calculated from the difference $F(b) - F(a)$

<u>*Example 1*</u>. Consider a normal distribution with average $\mu = 25$ and standard deviation $\sigma = 3$. Evaluate the probability corresponding to the interval between 20 and 29.First of all we shall calculate the two relevant values of z:

$$z_1 = \frac{x-\mu}{\sigma} = \frac{20-25}{3} = -1.67 \qquad z_2 = \frac{x-\mu}{\sigma} = \frac{29-25}{3} = 1.33$$

We can evaluate in table 5.6 the cumulative probability $F(z)$. Being z_1 negative, we shall evaluate F(-1.67)=1 - F(1.67) and from the table (17th line, 8th column) we get F(1.67)=0.9525, therefore F(-1.67)=0.0475. For z_2 it is simpler: we get directly from the table (14th line, 4th column) F(1.33)= 0.9082. The probability to get x within the selected interval 20 to 29 is therefore F(1.33)- F(-1.67)=0.8607.

<u>Example 2.</u> A commonly required evaluation, being known the probability level (called *confidence level*), is to define the relevant interval of the variable (called *confidence interval*). We do this under an accessory condition of symmetry around μ. We want, therefore, to evaluate the values x_1 and x_2, symmetrical around μ, corresponding to a confidence level $\mathcal{P}$=95%, as shown in Fig. 5.21.

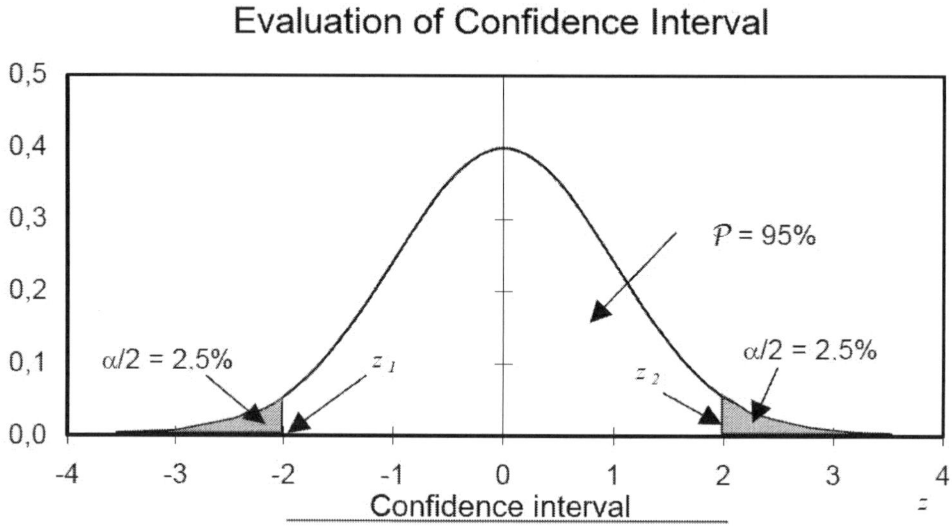

Fig. 5.21 Determination of the confidence interval (in the field of measurements, it corresponds to the uncertainty due to random factors)

First of all we shall evaluate the cumulative probabilities at x_1 and x_2. Being $\mathcal{P}$ the probability of the central part of the distribution, the probability of the outer parts, α (generally called *risk of statistical error of the first kind*), is its complement to 1:

$$\alpha = 1 - \mathcal{P} = 5\%$$

And, due to the symmetry, the probability of the left tail is:

$$\frac{\alpha}{2} = \frac{1-\mathcal{P}}{2} = 2.5\%$$

Always for symmetry, the cumulative probability from $-\infty$ to z_2 is:

$$1 - \frac{\alpha}{2} = 1 - \frac{1-\mathcal{P}}{2} = \frac{1+\mathcal{P}}{2} = 97.5\%$$

From table 5.6 the value 97.5% is in the 20th line, which has 1.9 as heading, and in the 7th column, which has 0.06 as heading, therefore $z_2 = 1.96$. Due to symmetry, the value $z_1 = -1.96$ is immediately obtained, but it could also be obtained as it is evident from the table that 2.5% shall have a negative value of z, and being $F(-z)=1 - F(z)$, that is 2.5% = 1 - $F(z)$ and $F(z)$ = 1 - 2.5% = 97.5%, therefore z = 1.96, as seen before, and $-z = z_1 = -1.96$. In case of measurement is conventional to consider at the confidence level of 95% the value of $z = \pm 2$

Now, from z_1 and z_2, considered in the measurement approximation, we can easily find x_1 and x_2 using equation 5.37:

$$x = \mu + z\sigma \tag{5.42}$$

inserting the values of z_1 and z_2 found before we get $x_1 = 25 - 2 \cdot 3 = 19$ and, for symmetry, or doing the calculation, $x_2 = 25 + 2 \cdot 3 = 31$.

The evaluation of the confidence interval is very important and is usually done in the same way, using, of course, different Density and Cumulative Functions for different statistical distributions whenever statistical tests are applied.

5.5.8 Student distribution

The confidence interval is very important for practical uses. If you have qualified your process with a variability index, you can test it. As distributions are frequently unbounded, establishing tolerances you have to accept that a percentage of your production is out. The starting point is therefore deciding an acceptable risk α, called statistical risk of error of the first kind, equal to the percentage of out of tolerance that you can accept. Thereafter, having the distribution and α, you can calculate the limits of confidence interval and check if really the production contains the expected acceptable percentage of pieces out of tolerance. As we have seen, in the frequent situation that your production is described by a normal distribution generally the standard deviation is used as dispersion index, but we have seen that the standard deviation is not a correct estimator, so, if you have evaluated your standard deviation with few degrees of freedom, it is likely that the percentage of pieces out of tolerance is higher than expected.

This is exactly what happened to W. S. Gosset, involved in the testing and research activity at the Guinness brewery in Dublin at the beginning of the twentieth century. Indeed, perhaps for the type of product, the beer, very interesting not only for doing tests, or to the widespread bad habit, unfortunately often practiced even today, to minimize any kind test, Gosset could only have a very limited number of samples for testing. Analysing the problem, Gosset found that, in case one could use only few samples, the distribution to be used was different from normal one. He published an article on this under the pseudonym of "Student", which explains why the statistical distribution proposed by him is so named.

This distribution has a bell shape similar to the normal distribution, and is asymptotic to it when the number n of the data used to calculate the standard deviation is high. But, when s is calculated with few data, the systematic error on s, and therefore the distribution shape, depends on their number, or better, by the number of degrees of freedom.

From the practical point of view, the use of Student distribution is quite similar to that of the normal distribution.

The standardized variable is called "Student's t" and the two limit values t_1 and t_2 are determined using the appropriate tables or functions, e.g. table 5.7 showing the value of Student's t for the most common values of cumulative probability and degrees of freedom.

The shapes of the distributions for $\nu = 3$ and $\nu = 6$ degrees of freedom are compared with that of the normal distribution in Fig. 5.22

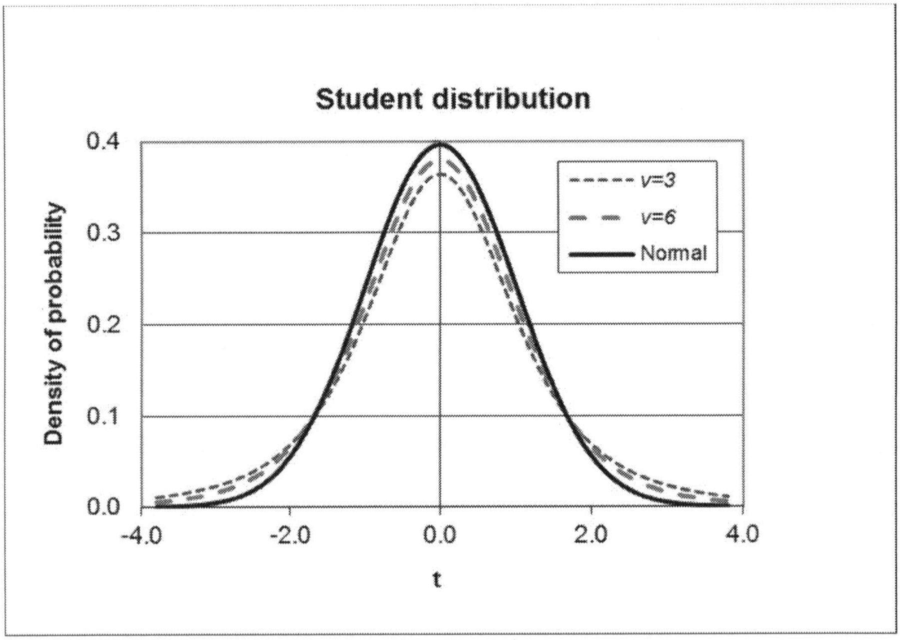

Fig. 5.22 Comparison of the normal distribution with Student's distribution for $\nu = 3$ and $\nu = 6$ degrees of freedom

To show the use of table 5.7 and compare the results obtained with normal distribution we repeat the evaluation of the confidence interval for the conventional confidence level of 95%, using $\nu = 4$ degrees of freedom, corresponding to a common industrial practice of 5 tests. As seen for the normal distribution, the cumulative probability at the upper level is 97.5%, so that from table 5.7 we get $t_2 = 2.78$. Being for symmetry $t_1 = -2.78$, the confidence interval will be between $x_1 = 16.66$ to $x_2 = 33.34$ really different from the values found with the normal $x_1 = 19$ to $x_2 = 31$. But if we do the same comparison with the conventional confidence level used in Quality, 99.7%, with the normal distribution there is a range of 6σ approximated by $6s$ from 16 to 34, with Student changes in the interval between 5.7 and 44.3, dramatically different!

*Tab. 5.7 Student's **t** values.*

Degrees of freedom	Cumulative probability								
	80%	85%	90%	93%	95%	97,5%	99%	99,85%	99,99%
1	1,38	1,96	3,08	4,47	6,31	12,71	31,82	212,19	3185,27
2	1,06	1,39	1,89	2,38	2,92	4,30	6,96	18,22	70,71
3	0,98	1,25	1,64	2,00	2,35	3,18	4,54	8,89	22,20
4	0,94	1,19	1,53	1,84	2,13	2,78	3,75	6,43	13,04
5	0,92	1,16	1,48	1,75	2,02	2,57	3,36	5,38	9,68
6	0,91	1,13	1,44	1,70	1,94	2,45	3,14	4,80	8,02
7	0,90	1,12	1,41	1,66	1,89	2,36	3,00	4,44	7,06
8	0,89	1,11	1,40	1,64	1,86	2,31	2,90	4,20	6,44
9	0,88	1,10	1,38	1,62	1,83	2,26	2,82	4,02	6,01
10	0,88	1,09	1,37	1,60	1,81	2,23	2,76	3,89	5,69
12	0,87	1,08	1,36	1,58	1,78	2,18	2,68	3,71	5,26
14	0,87	1,08	1,35	1,56	1,76	2,14	2,62	3,58	4,98
16	0,86	1,07	1,34	1,55	1,75	2,12	2,58	3,49	4,79
18	0,86	1,07	1,33	1,54	1,73	2,10	2,55	3,43	4,65
20	0,86	1,06	1,33	1,54	1,72	2,09	2,53	3,38	4,54
22	0,86	1,06	1,32	1,53	1,72	2,07	2,51	3,34	4,45
24	0,86	1,06	1,32	1,53	1,71	2,06	2,49	3,30	4,38
26	0,86	1,06	1,31	1,52	1,71	2,06	2,48	3,27	4,32
28	0,85	1,06	1,31	1,52	1,70	2,05	2,47	3,25	4,28
30	0,85	1,05	1,31	1,52	1,70	2,04	2,46	3,23	4,23
35	0,85	1,05	1,31	1,51	1,69	2,03	2,44	3,19	4,15
40	0,85	1,05	1,30	1,51	1,68	2,02	2,42	3,16	4,09
45	0,85	1,05	1,30	1,50	1,68	2,01	2,41	3,14	4,05
50	0,85	1,05	1,30	1,50	1,68	2,01	2,40	3,12	4,01
60	0,85	1,05	1,30	1,50	1,67	2,00	2,39	3,09	3,96
70	0,85	1,04	1,29	1,49	1,67	1,99	2,38	3,08	3,93
80	0,85	1,04	1,29	1,49	1,66	1,99	2,37	3,06	3,90
90	0,85	1,04	1,29	1,49	1,66	1,99	2,37	3,05	3,88
100	0,85	1,04	1,29	1,49	1,66	1,98	2,36	3,04	3,86

Note that Student's distribution provides the correction of the systematic error in the evaluation of s, but s has also a random variability, that is larger decreasing the number of degrees of freedom, therefore conclusions driven from tests with less than 15 degrees of freedom are really questionable.

5.5.9 Pearson's χ^2 distribution

As said before, the experimental value of the standard deviation s has a random variability. Pearson analysed the χ^2 distribution, related to the variability of s^2 around the theoretical value of σ^2 of a normally distributed population symbolized by the variable X with values x. That is, if from a population represented by the variable X normally distributed having a theoretical variance σ^2 we take different samples of n elements, for each sample i we have n values

x_{ij} (*i* indicates the sample, *j* the jth value of X within the ith sample), for each sample we can evaluate s_i^2 and consider a variable *V* having values v_i given by:

$$v_i = (n-1)\frac{s_i^2}{\sigma^2} = \sum_{j=1}^{n}\left(\frac{x_{ij}-m_i}{\sigma}\right)^2 \quad (5.43)$$

Using that distribution we can calculate the confidence interval of s^2. Let $\mathcal{P}$ be the confidence level. Using the χ^2, we evaluate the lower limit χ_1^2 and the upper limit χ_2^2 of the central zone corresponding to the probability $\mathcal{P}$ (in tab. 5.8 are reported the values of χ^2 corresponding to some useful levels of probability and degrees of freedom). We can say now that with the probability $\mathcal{P}$ values v_i are within the boundaries:

$$\chi_1^2 < v_i < \chi_2^2$$

And we replace v_i with what given in equation 5.43:

$$\chi_1^2 < (n-1)\frac{s_i^2}{\sigma^2} < \chi_2^2$$

thereafter:

$$\sigma^2 < (n-1)\frac{s_i^2}{\chi_1^2} \qquad (n-1)\frac{s_i^2}{\chi_2^2} < \sigma^2$$

Therefore we get the possible value of σ^2 around any evaluated value s_i^2:

$$(n-1)\frac{s_i^2}{\chi_2^2} < \sigma^2 < (n-1)\frac{s_i^2}{\chi_1^2} \quad (5.44)$$

For example, in our case we have evaluated a standard deviation of 3, that is a variance $s^2 = 9$ obtained with $n = 5$ data. We want to evaluate the confidence interval where the theoretical value σ^2 can be with the usual confidence level of 95%. As before, we determine the cumulative probability of 2.5% for the lower limit χ_1^2 and 97.5% for the upper limit χ_2^2. Going within the table 5.8 in the line of 4 degrees of freedom we get $\chi_1^2 = 0.48$ and $\chi_2^2 = 11.14$. Using equation 5.44 we have:

$$4\frac{9}{11.14} = 3.2 < \sigma^2 < 4\frac{9}{0.48} = 75$$

That is, our standard deviation, calculated to be 3, could be between 1.8 and 8.7. This will give you realistic information about the effects of random variability and of the importance of using a sufficient number of data. You can check that using 15 data, this variability is reduced between 2.2 and 4.7, a really significant variation, but acceptable in some cases. To have a random effect in the range of 20% on the standard deviation it is necessary to use about 50 data. This allows us to understand why the GUM highlights how the uncertainty itself is strongly "uncertain" and states the opportunity to declare it with one or at most two significant digits and, finally, urge to use the contributions based on information and past experience more than those determined by statistics with few degrees of freedom.

Tab. 5.8 Values of χ^2

Degrees of freedom	Cumulative Probability									
	1,0%	2,5%	5,0%	7,50%	10,0%	90,0%	92,5%	95%	97,5%	99,0%
1	0,00	0,00	0,00	0,01	0,02	2,71	3,17	3,84	5,02	6,63
2	0,02	0,05	0,10	0,16	0,21	4,61	5,18	5,99	7,38	9,21
3	0,11	0,22	0,35	0,47	0,58	6,25	6,90	7,81	9,35	11,34
4	0,30	0,48	0,71	0,90	1,06	7,78	8,50	9,49	11,14	13,28
5	0,55	0,83	1,15	1,39	1,61	9,24	10,01	11,07	12,83	15,09
6	0,87	1,24	1,64	1,94	2,20	10,64	11,47	12,59	14,45	16,81
7	1,24	1,69	2,17	2,53	2,83	12,02	12,88	14,07	16,01	18,48
8	1,65	2,18	2,73	3,14	3,49	13,36	14,27	15,51	17,53	20,09
9	2,09	2,70	3,33	3,78	4,17	14,68	15,63	16,92	19,02	21,67
10	2,56	3,25	3,94	4,45	4,87	15,99	16,97	18,31	20,48	23,21
12	3,57	4,40	5,23	5,82	6,30	18,55	19,60	21,03	23,34	26,22
14	4,66	5,63	6,57	7,24	7,79	21,06	22,18	23,68	26,12	29,14
16	5,81	6,91	7,96	8,71	9,31	23,54	24,72	26,30	28,85	32,00
18	7,01	8,23	9,39	10,21	10,86	25,99	27,22	28,87	31,53	34,81
20	8,26	9,59	10,85	11,73	12,44	28,41	29,69	31,41	34,17	37,57
22	9,54	10,98	12,34	13,28	14,04	30,81	32,14	33,92	36,78	40,29
24	10,86	12,40	13,85	14,85	15,66	33,20	34,57	36,42	39,36	42,98
26	12,20	13,84	15,38	16,44	17,29	35,56	36,98	38,89	41,92	45,64
28	13,56	15,31	16,93	18,05	18,94	37,92	39,38	41,34	44,46	48,28
30	14,95	16,79	18,49	19,66	20,60	40,26	41,76	43,77	46,98	50,89
35	18,51	20,57	22,47	23,76	24,80	46,06	47,66	49,80	53,20	57,34
40	22,16	24,43	26,51	27,93	29,05	51,81	53,50	55,76	59,34	63,69
45	25,90	28,37	30,61	32,14	33,35	57,51	59,29	61,66	65,41	69,96
50	29,71	32,36	34,76	36,40	37,69	63,17	65,03	67,50	71,42	76,15
60	37,48	40,48	43,19	45,02	46,46	74,40	76,41	79,08	83,30	88,38
70	45,44	48,76	51,74	53,75	55,33	85,53	87,68	90,53	95,02	100,43
80	53,54	57,15	60,39	62,57	64,28	96,58	98,86	101,88	106,63	112,33
90	61,75	65,65	69,13	71,46	73,29	107,57	109,97	113,15	118,14	124,12
100	70,06	74,22	77,93	80,41	82,36	118,50	121,02	124,34	129,56	135,81

Another common use of the Pearson's distribution allows to check if your experimental data, statistically described with an histogram, are suitable for being represented by a theoretical distribution. This test is called chi-square test, frequently written with the Greek letter χ^2. It is often used as a test of normality, and, more in general, χ^2 test is used for comparing experimental and theoretical distributions. This is important in many cases, for the help that one get validating, with a not so large number of data, a hypothesis of distribution for the phenomenon under evaluation.

The usual way χ^2 test is done consists in grouping experimental data as in a histogram and comparing their absolute frequency with the relevant expected values obtained with the theoretical distribution.

The field of the n experimental values x_i or that of the significant region of the statistical theoretical distribution used for comparison is subdivided in a number N of intervals, frequently called categories.

One has to consider two conditions: if N is large the description by the histogram is more detailed, but the number of data in each category is smaller, therefore the information driven from the category is not robust. The dimension of the categories should be such that each contains at least 3 data.

Just as indication one can try with $N \approx \sqrt{n}$ to cover the range of data, adding the necessary categories to cover the significant part of the theoretical distribution. It is also possible to use intervals larger where distributions have lower densities or to adopt a quantiles subdivision. Categories are therefore contiguous and subdivided by boundary values $c_0 \ldots c_N$.

For each category, the absolute experimental frequencies fa_j (number of data contained in the jth category), and the absolute theoretical frequencies ft_j shall be evaluated.

For ft_j it is necessary to evaluate the relative theoretical frequencies from the density of probability function $\mathrm{f}(x)$ of the theoretical distribution. You evaluate the theoretical probability of each category using the tables or the function of the relevant cumulative probability $\mathrm{F}(x)$.

The absolute theoretical frequency is obtained multiplying the relative theoretical frequency for the number of data n:

$$ft_j = n \cdot \left[\mathrm{F}(c_j) - \mathrm{F}(c_{j-1})\right]$$

It is now possible to use a variable, W, which value can easily be calculated:

$$w = \sum_{j=1}^{N} \frac{(fa_j - ft_j)^2}{ft_j} \qquad (5.45)$$

It can be demonstrated that W has a χ^2 distribution with N-3 degrees of freedom (for each of the N categories there is a possible difference, but there are three constraints, because usually the theoretical distribution is identified by an average and a standard deviation equals to that of the experimental data, and the histogram is done so that its total area is equal to 1, the area of the theoretical distribution).

Knowing that our calculated value w should arrive from a χ^2 distribution, we can decide how much is the maximum acceptable risk of error of the first kind and, thereafter, we can evaluate the confidence interval.

If the calculated w is outside that interval we reject the hypothesis that our data arrive from the theoretical distribution considered, and doing this we are aware that this assessment is false with a probability corresponding to the accepted risk of error. Usually the assessment is formulated with wording similar to this: "Being w outside the confidence interval, we can say, with a risk of error α, that the hypothesis that the experimental data are distributed following the theoretical distribution considered is not acceptable".

On the contrary, when w is inside the confidence interval, no conclusion can be drawn.

Usually the assessment is simply formulated: “Being w inside the confidence interval, there is no reason to reject the hypothesis that the experimental data are distributed following the theoretical distribution considered”.

5.5.10 Fisher F distribution

An important observation was made by R. A. Fisher on the benefits of the evaluation of the ratio of the variances. Consider, for instance, a testing room with r different operators. We will know if the difference of the test results is due to possible systematic differences between the operators.

We ask all operators to do repeated q times measurements of the same measurand and we take the average of the results of each operator as the best available indicator for evidencing his bias.

The dispersion of all the results obtained can be due to systematic differences among operators, or to random effects due to various influence quantities. Now it is possible to separate these two factors: we can evaluate the variance of our data in a way that the systematic differences among operators do not affect the result or, on the other side, in a way that feels completely the effects of suspected influence quantities, therefore giving a good indicator of their effect.

For the first way, if we evaluate the variance of the data of the first operator, this do not involves data of other operators, therefore do not feel any effect of the differences from operators.

An equivalent action can be done with data of all others operators, so that we have r independent values of the variance of our data and we got a more robust result taking their average. It was considered that in this way data were subdivided in r homogeneous groups (homogeneous because do not contain what a factor suspect of producing systematic differences), therefore the variance obtained was called *variance within the groups* and indicated with the symbol s_W^2.

For the second way, it is also possible to evaluate the variance of all the data evidencing possible differences between the operators; these differences are certainly contained in the averages of the different operators, therefore consists in evaluating the variance of the averages.

To pass to the variance of individual data we can use equation 5.17 and multiply the variance of the average for q, which is the number of averaged data.

This variance contains the possible systematic differences between the groups and therefore is called *variance between the groups* and indicated with the symbol s_B^2.

If there are no systematic differences among operators, these two variance estimations should have equal expected values, that is their ratio is expected to be 1, but for random effects. Fisher studied the statistical distribution of the ratio of the variance estimation, considering the variable given in equation 5.46.

$$F_{m,n} = \frac{\dfrac{s_1^2}{\sigma_1^2}}{\dfrac{s_2^2}{\sigma_2^2}} = \frac{\dfrac{1}{\sigma_1^2}\dfrac{\sum_{i=1}^{m+1}(x_{1i}-m_{1x})^2}{m}}{\dfrac{1}{\sigma_2^2}\dfrac{\sum_{j=1}^{n+1}(x_{2j}-m_{2x})^2}{n}} \qquad (5.46)$$

where $\boldsymbol{m}$ are the degrees of freedom of the variance at the numerator and $\boldsymbol{n}$ of the variance at the denominator. In our simple case we evaluated in two ways the same variance, therefore s_W^2 and s_B^2 are estimators of the same variance, therefore $\sigma_1^2 = \sigma_2^2$.

$F_{m,n}$ is the variable of a statistical distribution called Fisher's *F*, and its values depend on $\boldsymbol{m}$, $\boldsymbol{n}$ and the cumulated probability, therefore cannot be given in a single table, but shall use some tables each one for a given value of cumulative probability, as given in table 5.9.

Tab. 5.9 Fisher's F values.

Cumulative probability 90%										
n \ m	2	4	6	10	15	20	30	40	60	100
2	9.00	9.24	9.33	9.39	9.42	9.44	9.46	9.47	9.47	9.48
4	4.32	4.11	4.01	3.92	3.87	3.84	3.82	3.80	3.79	3.78
6	3.46	3.18	3.05	2.94	2.87	2.84	2.80	2.78	2.76	2.75
10	2.92	2.61	2.46	2.32	2.24	2.20	2.16	2.13	2.11	2.09
15	2.70	2.36	2.21	2.06	1.97	1.92	1.87	1.85	1.82	1.79
20	2.59	2.25	2.09	1.94	1.84	1.79	1.74	1.71	1.68	1.65
30	2.49	2.14	1.98	1.82	1.72	1.67	1.61	1.57	1.54	1.51
40	2.44	2.09	1.93	1.76	1.66	1.61	1.54	1.51	1.47	1.43
60	2.39	2.04	1.87	1.71	1.60	1.54	1.48	1.44	1.40	1.36
100	2.36	2.00	1.83	1.66	1.56	1.49	1.42	1.38	1.34	1.29

Cumulative probability 95%										
n \ m	2	4	6	10	15	20	30	40	60	100
2	19.00	19.25	19.33	19.40	19.43	19.45	19.46	19.47	19.48	19.49
4	6.94	6.39	6.16	5.96	5.86	5.80	5.75	5.72	5.69	5.66
6	5.14	4.53	4.28	4.06	3.94	3.87	3.81	3.77	3.74	3.71
10	4.10	3.48	3.22	2.98	2.85	2.77	2.70	2.66	2.62	2.59
15	3.68	3.06	2.79	2.54	2.40	2.33	2.25	2.20	2.16	2.12
20	3.49	2.87	2.60	2.35	2.20	2.12	2.04	1.99	1.95	1.91
30	3.32	2.69	2.42	2.16	2.01	1.93	1.84	1.79	1.74	1.70
40	3.23	2.61	2.34	2.08	1.92	1.84	1.74	1.69	1.64	1.59
60	3.15	2.53	2.25	1.99	1.84	1.75	1.65	1.59	1.53	1.48
100	3.09	2.46	2.19	1.93	1.77	1.68	1.57	1.52	1.45	1.39

Cumulative probability 90%										
n \ m	2	4	6	10	15	20	30	40	60	100
2	39.00	39.25	39.33	39.40	39.43	39.45	39.46	39.47	39.48	39.49
4	10.65	9.60	9.20	8.84	8.66	8.56	8.46	8.41	8.36	8.32
6	7.26	6.23	5.82	5.46	5.27	5.17	5.07	5.01	4.96	4.92
10	5.46	4.47	4.07	3.72	3.52	3.42	3.31	3.26	3.20	3.15
15	4.77	3.80	3.41	3.06	2.86	2.76	2.64	2.59	2.52	2.47
20	4.46	3.51	3.13	2.77	2.57	2.46	2.35	2.29	2.22	2.17
30	4.18	3.25	2.87	2.51	2.31	2.20	2.07	2.01	1.94	1.88
40	4.05	3.13	2.74	2.39	2.18	2.07	1.94	1.88	1.80	1.74
60	3.93	3.01	2.63	2.27	2.06	1.94	1.82	1.74	1.67	1.60
100	39.00	39.25	39.33	39.40	39.43	39.45	39.46	39.47	39.48	39.49

5.5.11 General uses of statistical distributions

Statistics allows getting answers with statistical tests that have a general working hypothesis. If we define a given situation (pieces produced by a defined machine tool, measurement results obtained by an identified instrument, or even more complex, as measurement obtained in a given testing room, working over 24 hours, in different ambient conditions, with six operators), the outcomes of that situation can be described by a statistical distribution. If we do a very large number of tests we can, in principle, identify the distribution that describes the said situation.

The first step, very reasonable if we remember all what said on functional specification, is to pose clearly the question in a statistical way. Having a statistical distribution and an outcome, statistics can say if it is not likely to have such level of outcome from the given distribution. For instance consider measurement uncertainty as defined in GUM: "*parameter, associated with the result of a measurement, that characterizes the dispersion of the values that could reasonably be attributed to the measurand*". The statistical translation is immediate: given the statistical distribution of measurement results, uncertainty is the confidence interval evaluated at a reasonable confidence level. Therefore, we shall translate "result within the uncertainty" in "outcome pertaining to the selected distribution" and this is what in statistics is called *null hypothesis*. Statistical test can give only one real answer: *reject the null hypothesis*. The complementary answer that could seem logical, acceptance of the null hypothesis is not attainable in the same way, in fact by measurement you can never arrive to the conclusion that two measurands are equal. The complementary answer is: *the null hypothesis cannot be rejected*, which is an answer declaring ignorance.

For uncertainty the null hypothesis is: the measurement result examined is an outcome of the distribution defining the measurement conditions.

The second step can be well understood considering the example of uncertainty, and specifically the uncertainty on the strength of a batch of bolts: we will provide the costumer with reasonable information on the lower strength of the bolts of the batch. Now, what is reasonable? With statistical tests we can give the answer at a given confidence level, that is, looking at Fig. 5.21 we established the confidence interval at a confidence level of 95%, that is the interval containing 95% of the outcomes of the distribution. If we take that interval as uncertainty interval, we take the lower limit of that interval as lower limit of the uncertainty; but there is 2.5% of possible outcomes lower than that lower limit. This means that, declaring the lower limit of strength as that lower limit of the confidence interval, we have a risk to get 2.5% of the bolts of the batch that have a strength lower than the lower limit declared, that is our declaration has a 2.5% risk to be false (called risk of error of the first kind on the lower limit). We have the same at the upper side (called risk of error of the first kind on the upper limit). A reasonable interval means that the risks of error involved are acceptable, therefore the second step to get a significant answer from statistical tests is to establish acceptable risks of error. How can we do this? In the same way it is usually done, looking at consequences. If the bolts will be used for the coachwork and one bolt brakes, the consequences are a lightly noisy condition, while if bolts are used to take the wheels at their place and one bolt brakes, consequences can be very dangerous. It is evident that in this second case it is reasonable to accept a very small risk, while in the first case the risk can be much higher. It is evident that smaller risks involve higher costs, moreover we will see soon then there is another type of risks, the risk of error of the second kind, which increases as that of the first kind decreases, therefore it is not possible just to establish in any case the minimum risk of the first kind. Note, also, that as the consequences can be different on the upper or lower side, also the relevant risks should be different. In our example if the strength of a bolt is lower than the lower limit of the declared uncertainty, the consequences can be that the bolt will break, but the breaking in this condition is slow. If the strength of a bolt is higher than a technological limit, it is likely that the bolt material is brittle, producing an abrupt breaking, very dangerous. In that case the upper risk shall be lower than the lower risk. When no such differences are involved a total risk of error α is considered, equally subdivided in lower and upper parts.

Now having the statistical distribution related to the null hypothesis and the risks of error, we can apply calculations with tables or statistical functions of the relevant distribution, to get from the values of risks the corresponding value of the distribution variable. We define in this way the confidence interval and we can compare with it the outcome under examination. The answer of the statistical test can be:

- "NO"; the value of your result is outside the confidence interval, therefore it is not justified, at the given confidence level, by the random variability corresponding to the statistical distribution adopted: there is something else than random errors, a systematic error or something outside the statistics given;
- "I don't know"; the value of your result is inside the confidence interval, therefore is justified, at the given confidence level, by the random variability corresponding to the statistical distribution adopted: this will not involve the fact that there are no systematic errors or something outside the statistics given, but only that their effect is confused within the random variability.

Statistical inference

Frequently the procedure involved consists in qualifying our measurement system. The part of qualification referred to random effects consists on taking a number of repeated measurements to determine the distribution of results. We can consider the average x_m of our results as the value of the measurand, therefore our distribution gives us the information of the confidence interval, defined by the distances d_L and d_U from the average, at a given confidence level $\mathcal{P}_c$. Thereafter we use our measurement system to do other measurement, taking care not to change the measuring conditions, therefore to keep for the new measurement the same distribution identified in the qualification phase. So we get a new value x and we know that with the probability $\mathcal{P}_c$ it is within the confidence interval around the qualification average x_m, our best information of the measurand value. By statistical inference we can say that the qualification average x_m, with a probability $\mathcal{P}_c$, is an identical confidence interval around x.

In fact, being with a probability $\mathcal{P}_c$:

$$x_m - d_L < x < x_m + d_U$$

we get the minimum value of x_m down of x:

$$x_m - d_L < x \Rightarrow x_m < x + d_L$$

and the maximum value of x_m up of x:

$$x_m + d_U > x \Rightarrow x_m > x - d_U$$

Therefore, with the same probability $\mathcal{P}_c$ we have:

$$x - d_U < x_m < x + d_L \tag{5.47}$$

Note that the interval is the same as the confidence interval, but a possible asymmetry due to a difference between d_L and d_U is reversed.

Confidence level and risks of error of first and second kind

We have seen that confidence interval is in the central part of the distribution, corresponding to an interval of the distribution variable in which the probability of outcomes $\mathcal{P}_c$ is the confidence level. Note that this concept has embedded a condition of symmetry around the average. The complement to 1 of the confidence interval is the risk of error of the first kind α, again with an intrinsic condition of symmetry between the risk of error of the first kind on the lower part $\alpha_L = \alpha/2$ and the upper part $\alpha_U = \alpha/2$. The two risks of error of the first kind can be different, as the consequences of being out in the lower or upper parts can be different, and in that cases is perfectly possible to apply the given concepts end frequently also the equations.

It is important that always the two types of risk of error are taken into account. In fact consider a statistical test (or a measurement comparison between two measurands). The null hypothesis can be true or false and the answer of the test can be a rejection or no reasons for rejection. We can see this in Fig. 5.23:

Risk of error		Null Hypothesis	
		True	False
Test answer	Rejection	First Kind Error	CORRECT
	No rejection	CORRECT	Second Kind Error

Fig. 5.23 Possible test answers and relevant errors. When the hypothesis is true and the test gives a rejection the error of first kind is involved, while when the hypothesis is false and the test gives no-rejection is the case of the error of second kind.

Your null hypothesis can be true, but you get a data in a tail, outside the confidence interval, therefore your test will not accept the null hypothesis. This can happen exactly in the percentage of data out of the confidence interval, that is with a risk $\alpha = 1 - \mathcal{P}_c$ (risk of error of the first kind). The amount of this risk of error shall be decided before considering the consequences and the confidence interval is evaluated on its base, therefore is completely under control: yourself decide how much it shall be.

For describing the risk of second kind, we shall consider the distribution **A** describing the main phenomenon involved and a different distribution **B** describing the effect of some systematic changes; for instance a measurement made in significantly different ambient conditions. This is shown in Fig. 5.24 with the distribution **A** and the upper limit of the confidence interval at $x = 21$ and distribution **B** slightly different.

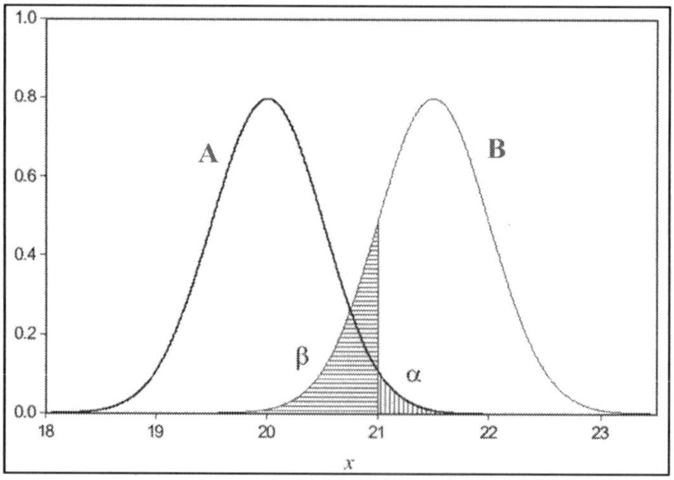

Fig. 5.24 Graphical presentation of the errors of first and second kind.

It is really possible that an outcome of distribution **B** enters in the confidence interval of distribution **A**, so that our test sees a result within the confidence interval and do not reject the null hypothesis. The probability of this event is given graphically by the area β that is the risk of error of the second kind. Being unknown the position and the shape of distribution **B**, β cannot be evaluated, therefore the error of second kind is not under control. Nevertheless it is easy to see in Fig. 5.24 that if you can accept to increase the risk of first kind, that is the area α, the boundary of the confidence interval moves toward the left and the area β, representing the risk of error of the second kind, decreases. Therefore it is important to adopt the maximum acceptable risk of error of the first kind.

6 STATISTICAL MANAGEMENT OF EXPERIMENTAL DATA

6.1 INTRODUCTION

The description of the behaviour of a measurement complex has highlighted three fundamental metrological characteristics: repeatability, reproducibility and trueness, strongly connected to the random and systematic effects. Measurement operations produce what is frequently called a cloud of values. The three characteristics listed above, have been defined to handle these results. The interest of the user of the measurement complex is to extract from such cloud the value to be assigned to the measurand and the uncertainty connected to it. We can underline that the measures obtained should be considered *distorted*, due to the presence of systematic effects, and *hazy*, for the presence of random ones. Moreover, it must not be forgotten that it is always possible that there were actual measurement accidents.

The purpose of the analysis of the experimental results, made with statistical methods, should locate from the cloud of values obtained, a *position index*, that allows to isolate the value to be assigned to the measurand and a *dispersion index*, that allows to estimate the uncertainty. We also need to implement those techniques designed to detect the presence of systematic factors and possible accidents. The identification of such disturbing factors must be done by reviewing the operating conditions and the basic concepts according to whom the measuring procedure is organized. These details are related to a specific experience of the sector in which the measurement takes place, therefore shall be discussed with the process experts.

The development of general principles that lead to highlight the presence of undesirable factors and, after their elimination to determine the position index and the dispersion index, allows to manage the experimental data in an appropriate manner to the Standard Specification requests. These could be considered fulfilled by the determination of the average value and standard deviation, but, as seen before, the presence of systematic factors or measurement (or production in case of products) accidents may lead to misleading differences.

It shall be clear that the analysis can only indicate systematic effects and possible accidents. This is a good help for the technical part, which, taking the data analysis indications, can understand the technical problems involved and decide how to avoid the errors and how to manage the outliers.

6.2 THE DESCRIPTION OF EXPERIMENTAL DATA

To present a typical approach we use the data in Table 6.1, on which will be carried out the entire process of statistical analysis. Data collected are usually placed in a table, but it is certainly hard to identify in a table of numbers possible tendencies or irregularities or strange data. To make the situation more visible it is appropriate to adopt a graphic presentation, which consents to realize more concretely the situation. The first operation to do is to count the experimental data. For this purpose there are, in spreadsheet, already prepared functions.

6.2.1 The counting of the experimental data

The table of raw data, together with all the information that might allow the repetition of the experiment, must be maintained in safe conditions, avoiding any type of correction and modification, and included in the final report. With the use of modern instruments or computers there is the real risk of modifying the starting situation and loose the original information. This is evident with computers, if you modify some data and save the file, it could be impossible to restore the original data. With modern instruments performing automatically some operation, some data can never be available, for example, the use of instruments with the automatic zero, that provide directly the differences from the starting value any information on the position of the zero is not collected.

Tab. 6.1 Experimental data obtained with the length measurement of a workpiece made by five different operators. The measurements are expressed in millimetres.

Oper. 1	Oper. 2	Oper. 3	Oper. 4	Oper. 5
8.0038	8.0040	8.0060	8.0092	8.0053
8.0051	8.0025	8.0041	8.0104	8.0076
8.0037	8.0093	8.0028	8.0110	8.0113
8.0017	8.0025	8.0049	8.0087	8.0112
8.0023	8.0046	8.0025	8.0047	8.0066
8.0140	8.0020	8.0030	8.0076	8.0065
8.0080	8.0046	8.0073	8.0056	8.0114
8.0058	8.0085	8.0065	8.0113	8.0064
8.0031	8.0056	8.0071	8.0068	8.0065
8.0084	8.0037	8.0064	8.0093	8.0057

In case the acquisition of experimental data is not automatically done it is necessary to manage the data manually.

For counting of the data manually and also to bring them easily in ascending order (to find the maximum and minimum) it is useful to arrange an axis, on some lines if necessary, and to put on it a check sign for each data of your table, as represented in Fig. 6.1. To place them on an axis you need to leave a certain space in the preceding and subsequent lines respect the one of the axis, to ensure that, where necessary, it can be extended both towards smaller values and towards higher ones. The starting point, therefore, is not very important.

For convenience, the constant part of data is not reported on the scale of the axis, in our case 8.000 mm is taken away, so that the scale is given in micrometres: from the first data it will remain 3.8 μm, from the second 5.1 μm and so on. The axis unit of format is equal to the measurement resolution.

```
               x                                       x
               x           x        x         x       xx          x
       x  x  x x  x xx     xx xx    xx x x x  xxx x  xxxx x  x x  x
|....,....|....,....|....,....|....,....|....,....|....,....|....,....|....,....
1         2         3         4         5         6         7

             x                   x
x   xx x    xx          x     x xxx                         x
|....,....|....,....|....,....|....,....|....,....|....,....|....,....|....,....
8         9         10        11        12        13        14
```

Fig. 6.1 Graphic of the counting of the experimental data of table 6.1

The data contained in the table are taken in succession giving each a sign, for example an x, in correspondence of its value marked on the axis. If in the place corresponding to a certain value there are already one or more signs, we overlap the last one to the previous, as represented in Fig. 6.1.

6.2.2 The presentation of the experimental data

The diagram shown in Fig. 6.1, typically called *dot-plot*, represents the shape of the experimental data distribution, but in general it is not easily readable because it is too minutely divided and it is often represented on two or more rows. A better representation is the histogram[1] as said in chapter 5. We apply it to our data. In the example considered we can take intervals of 1.5 μm, starting from 1.45 μm. The categories are then defined by the interval 1.45 - 2.95 - 4.45 and so on.

From the examination of Fig. 6.1 the absolute frequency f_a can easily be evaluated by counting the number of data in each category. The first category contains 7 data, the second 7 data, the third one 10, etc., as shown in column f_a of Table 6.2.

The graphical representation can be made with a bar chart (Fig. 6.2a) whose heights represent the absolute frequencies, but in this way will not be comparable with the distributions of probability[2]. To obtain this comparison we must, first of all, estimate the probability corresponding to each category width. This is done by calculating the relative frequency given by the ratio between the number of data that fall into a certain category (selected event) and the total

---

[1] Column chart and histogram are not synonymous. The histogram is a specialized form of column chart characterized by the fact that the quantity that each column represents is not expressed by the height of the column, but by its area.

[2] The probability distributions are characterized by the fact that the probability that a given value will be in the range between a and b is represented by the area under the probability distribution. The experimental analogous of probability is the relative frequency.

number of data (sample space). The values in the relative frequency f_r column are calculated in this way.

Tab. 6.2 Distribution of the experimental frequency.

From [mm]	To [mm]	Central value [mm]	f_a	f_r	ρ_f [mm^{-1}]
8.00145	8.00295	8.0022	7	0.140	93.33
8.00295	8.00445	8.0037	7	0.140	93.33
8.00445	8.00595	8.0052	10	0.200	133.33
8.00595	8.00745	8.0067	10	0.220	146.67
8.00745	8.00895	8.0082	6	0.120	80.00
8.00895	8.01045	8.0097	4	0.080	53.33
8.01045	8.01195	8.0112	5	0.100	66.67
8.01195	8.01345	8.0127	0	0.000	0.00
8.01345	8.01495	8.0142	1	0.020	13.33

To complete the possibility of comparison we must represent the relative frequencies with a histogram (Fig. 6.2b) in which the relative frequency is represented by the area of the column.

To do this, being known the base of the columns (category width) and its area (relative frequency) is useful to calculate the height, which takes the name of *frequency density* ρ_f, dividing the values of relative frequency for the width of the category. This way the trend of the examined phenomenon is presented in its overall shape.

This presentation is not overly sensitive to the amplitude and position of the categories.

We must try now to extract from the experimental data their information content, in particular a position (*position index*) and dispersion (*dispersion index*) of the experimental data.

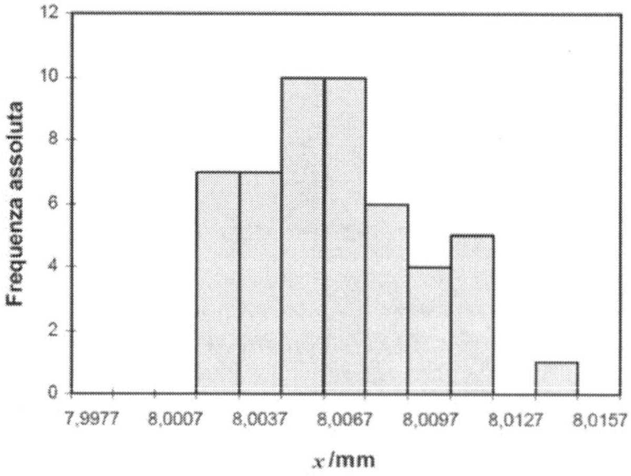

Fig. 6.2a Trend of the absolute frequencies.

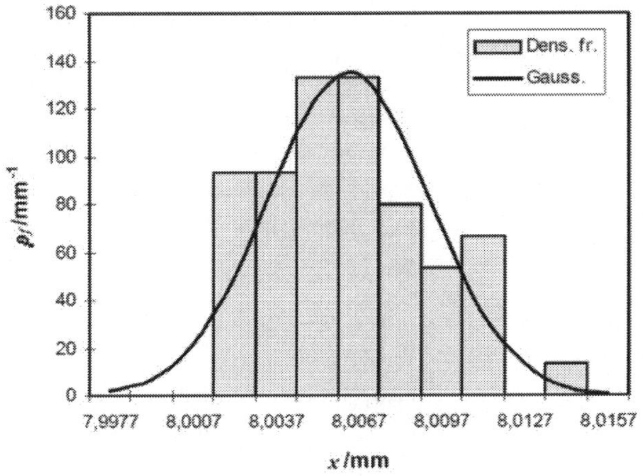

Fig. 6.2b Histogram of the relative frequency. The frequency density ρ_f is just a variable of convenience to draw the corresponding column chart. The histogram presents the relative frequency as the area of the columns. It is, therefore, comparable with the statistical distributions, since the whole area, corresponding to the total of the relative frequencies, is 1.

6.2.3 Determination of central tendency and dispersion

Many phenomena are described by a range of values rather than a single one. For an easier management of the comparisons, we try to extract from the group of values obtained in the experimental tests what is believed to be the value more significant for the description of the examined phenomenon. It appears separately, then, an interval as information on the dispersion, often represented in a conventional manner.

In section 5.3 it has already been introduced, in general, the method of evaluation of position index and dispersion index and they have been given the formulas repeated here, for the reader convenience:

$$m_x = \frac{1}{n}\sum_{i=1}^{n} x_i \tag{6.1}$$

$$s_x^2 = \frac{\sum_{i=1}^{n}(x_i - m_x)^2}{n-1} \tag{6.2}$$

In the management of experimental data, however, we cannot always recourse to a theoretical distribution, so it is useful to use other methods exemplified here.

6.2.3.1 The evaluation of the position index

To obtain an estimate of the most relevant value of the measurand one must process the measurement data. As seen in section 5.4, this can be done with different methods. For the analysis of the experimental data the most frequently used methods, in addition to the arithmetic average, cited above, are based on the determination of the *median* and *mode*. For example, the median of the 5 data of Table 6.3 is obtained by putting them in order and taking the central value: 8.0053. In case of 6 data, the position index is between the third and fourth data of the ordered series: 8.0056; as described in Table 6.4. For the data of Table 6.1, the median can be found using the sorting diagram of Fig. 6.1. Since 50 values have been collected, the median is the midpoint between the twenty-fifth and twenty-sixth value: 8.00615 mm.

Tab. 6.3. Set of 5 measuring data and ordered series of the five values. The central one, in bold, is the median.

Data	8,0040	8,0038	8,0060	8,0092	8,0053
Sorted data	8.0038	8.0040	**8.0053**	8.0060	8.0092

Tab. 6.4. Set of 6 measurement data and their ordered set. The central point of the two central values, 8.0056, is the median.

Data	8,0040	8,0038	8,0073	8,0060	8,0092	8,0052
Sorted data	8.0038	8.0040	**8.0052**	**8.0060**	8.0073	8.0092

The second method for estimating the value of reference uses the concept of *mode*, which is the point of maximum frequency (value that appeared the highest number of times in the measurement). In case of the data of Table 6.1, it can be seen from Fig. 6.1 which both 8.0025 mm, 8.0065 mm are present three times. We could, therefore, say that there are two *modes*. This, however, is also due to the excessive segmentation of the data. A better regularity is obtained by the division into categories and with the assumption, as modes, of the more numerous categories: from 8.00445 mm to 8.00595 mm and from 8.00595 mm to 8.00745 mm.

It can be observed that the average, as the position index, has the most rigorous characteristics, in the sense that it is well framed in the discussion of the normal statistics. Compared to the median it has, however, an increased sensitivity to the presence of disturbing factors such as measurement accidents and systematic effects.

The determination of the position index is crucial for assessing the characteristic of trueness; this is because, as seen from the central limit theorem, the position index is able to decrease the effect of the variability of the individual data, giving more definite information[3]. The operation of determining

[3] As a consequence the number of significant digits with which the average value is written can be incremented by 1 compared to the one of the initial data.

this value is placed in a peculiar condition compared to the other operations to which measurement data are subject to calculate the magnitude of interest. In fact, this may lead to a reduction of uncertainty, related to the theorem of central limit, but limited to the effects of random factors. Basically, if you suspect that an uncertainty factor depends on random contributions, you can decrease its effect replicating the measures and taking their average value. The decrease of the standard deviation is always deduced from the central limit theorem (equation 5.17) and is: of a factor two if you take four measures, of a factor three if you take nine. This method is widely used in modern electronic instruments that, being very fast in the measurement process, mediate on hundreds or sometimes thousands of results, thus reducing drastically the uncertainty related to background noise. It is clear that if there is a constant systematic error, if, for example, the sensitivity of the instrument is 1% greater than its correct value, all the results will be enhanced by 1%, and to mediate a large number will not help. In case of a variable systematic factor, for example the effect of temperature in a not thermostatic chamber, the replication of the measurements and the averaging operation decreases the uncertainty, but only the part compatibly to the variation of temperature that has been felt with the replications of measurement. If you continue the measurements for periods of one or more days, taking the data randomly (i.e. not grouped at certain times) you could eliminate most of the effect of the variability of the day-night temperature. If, instead, you take many grouped measurements in a short period, in which the temperature has undergone a minimum change, it will not be available any information of the effect of temperature variation and the averaging operation cannot do anything.

6.2.3.2 The evaluation of the results' dispersion and of the uncertainty

In the field of the measures in addition to the most significant value of the measurement results obtained it is essential to determine also the measurement uncertainty. This describes the field around the reference value where it is reasonable to have measurement results, taking into account all factors, random or systematic.

The uncertainty is a key concept to allow the use of the information given by the measure. The only estimate of the reference value, in fact, does not allow to draw any conclusions, as seen in Chapter 1.

6.2.3.2.1 The field of variability as uncertainty range

The manual counting of Fig. 6.1 also allows to see directly the field of variability of the results obtained (*range*), which could be taken as field of uncertainty.

We should, however, immediately note that if we took other data, in addition to the fifty examined, the extension may increase because some data might fall outside the extension of the previous data (this happens, especially, when taken data are few), but never decrease. Range is therefore systematically smaller than the variability field, much smaller when you have few data.

Management of few data is very common in quality tests; this introduces a common practice to refer to the statistical distribution (due to Pearson[4]) of the ratio between the range and the standard deviation σ.

6.2.3.2.2 The variability defined as probability field

An alternative hypothesis of evaluation of a variability index of the experimental results more independent from the number *n* of the examined data sample might be the interval that contains a certain percentage *P* of the data. In fact, it is true that with the increase of *n* the range increases, but the probability of getting data in a given interval remains theoretically constant. Therefore, the interval containing a given percentage of data is more stable than the range.

This method corresponds to the use of quantiles (see section 5.4) and is very effective when a large number of data is available. For example from Fig. 6.1 we can easily derive the table 6.5. The median is obtained easily that is correspondent to the upper limit of D5, Δx = 6.15 μm, equal to the one of the second quartile.

Tab. 6.5 Table of the experimental distribution of the deciles of the experimental data of Table 6.1 obtained from Fig. 6.1. For convenience, here are given the differences from 8 mm expressed in micrometres.

Decile	D1	D2	D3	D4	D5	D6	D7	D8	D9	D10
from	1.70	2.50	3.70	4.60	5.45	6.15	6.55	7.60	8.95	11.1
to	2.50	3.70	4.60	5.45	6.15	6.55	7.60	8.95	11.1	14.0

In our case, for example, if you choose as uncertainty interval the field that contains 90% of the data, you must discard the 10% in the bilaterally symmetrical form, 5% (5 centiles) between the data of lesser value and 5% of the data of higher value. For our sample of 50 data, we take the interval obtained discarding 2.5 data in the lower zone (lower limit 8.0022) and 2.5 data at the upper zone (upper limit 8.0113). The interval that describes the uncertainty of 90% should be between 8.0022 and 8.0113. It is evident how this method is hardly practicable with few data and strongly dependent on the position of only the data of the extremities. With fifty values, which is a large number compared to the real possibility of replication of normal industrial measurements, everything depends on the position of the three latest data at the lower and upper limits.

A more concrete possibility to determine indexes that allow a more robust evaluation of the dispersion of the experimental data is obtained using the difference between the third and the first quartile (*interquartile range*, IQR).

---

[4] If the experimental data have a normal distribution, you can use the relationship $E(R)=d_2(n)\sigma$, calculated by Pearson, between the standard deviation σ of the population and the expected value $E(R)$ of the extension of *R*, from which it is estimated the standard deviation $s=R/d_2(n)$. The values of $d_2(n)$ are given as a function of the number n of data examined (Douglas C. Montgomery, Applied Statistics and Probability for Engineers, Fifth Edition, Chapter 15)

In this case, in fact, the position of the border is related to a large group of experimental data (50% of the data present in the central area), so it is less affected by small displacements on few data. A traditional indication that allows to evaluate the central position and the dispersion is called *indication of five values*: minimum, first quartile, median, third quartile and maximum. These values can be easily obtained from the counting of the data, and, in the case of the example of tab. 6.1, are: minimum 1.7, first quartile 4.1, median 6.15, third quartile 8.0, maximum 14.0. The corresponding graph is called *boxplot* (see also section 5.4).

6.2.3.2.3 The use of the normal distribution

To give a more robust evaluation is necessary to use the information contained in all the examined data and this can only be done basing on assumptions about their statistical distribution.

In case of experimental data a starting hypothesis is linked to the fact that has been paid a great attention to avoid systematic errors and measurement accidents, therefore we presume that only the accidental errors remain in our data. The dispersion of the results is then related to the combined effect of many small random causes that can shift the result as towards higher values, as towards lower values. As we have seen, this very common situation involves a normal distribution. It is, therefore, possible to identify the experimental distribution with a normal distribution, estimating the position index μ of the theoretical distribution with the arithmetic average m of experimental data and the variance σ^2 with the variance of the experimental distribution s^2.

6.2.4 Statistical inference for the determination of confidence intervals

The possibility of representing the experimental data with a theoretical statistical distribution brings tangible benefits. In fact, the uncertainty estimation obtained evaluating the interval of the theoretical distribution that contains a high percentage of the experimental data does not present the disadvantage of depending only by the few experimental data in the extreme positions.

In this case, in fact, all the experimental data contribute to the estimation of μ and σ.

The relevant interval can be determined with the widest statistical base allowed by the number of the available experimental data. Nevertheless we shall not forget the correlation uncertainty (chapter 1) that is the possible error taking the normal distribution as a model.

The way of proceeding, already described in theoretical form, is simple but must be well understood. The concepts involved are:

a. establish the acceptable statistical risk of error of first kind α, corresponding to the probability that the null hypothesis will not occur (in our case that a result will be outside the estimated range of uncertainty, while belonging to the considered distribution). Consider also if it is unilateral or bilateral and in this case if symmetrical or not;

b. consider the theoretical statistical distribution that describes the population of the experimental data and identify it using *m* and *s* of the experimental data as its μ and σ. According to the examined variable (individual results, averages of groups of results, results of individual differences, differences of averages, and, in general, functions of individual results or of averages of groups of results) and the number of data from which the parameters of the theoretical statistical distribution have been estimated the corresponding distribution must be used (normal for a lot of data, Student's for limited data);
c. determine the limits of the distribution variable (z, x, t, or other variables related to the type of distribution), from the theoretical statistical distribution, the value and the risk of error modality (unilateral or bilateral), corresponding to the risk at lower level α_L and upper level α_U error.

As strongly stressed the risk of error of first kind shall be determined considering the consequences involved, but this is frequently not possible (at testing arrive samples without the indication of their use) therefore one shall use the conventional bilateral symmetrical condition with a confidence interval of 95% (risks of error 2.5%). In this case the conventional value will be declared and the effective one decided by the user.

We shall remember that we are working on the hypothesis of normal (or Student) distribution, which is likely to be only if data are free of systematic errors and measurement accidents; therefore, before establishing the theoretical confidence interval is necessary to perform a careful data analysis to clean data from unwanted disturbances.

6.3 COMPLETE PROCEDURE OF THE EXPERIMENTAL DATA ANALYSIS

Data analysis is really important for the industrial activity and in the framework of scientific research, for the detection of systematic factors.

In fact at industrial level the identification of a systematic factor within a production process can allow to apply correction, compensation or to avoid the causes producing the factor itself and get efficiently, in this way, a better production.

For scientific research systematic factors allow to understand in a better way the phenomenon under study, and produce a better description of it.

A generally efficient path for developing data analysis is as follows:

a. preliminary descriptive evaluation. As said before, a table of numbers usually does not show specific effects. A better vision of what is in the data is given by graphical presentation. For instance specific statistical software as a first step of the analysis give all the graphs of all the variables one versus each of the others (if you have one dependent variable and two independent one, they produce a triangular matrix half of a square 3x3 matrix, that is 6 small graphs). The result is better if data are modified to eliminate the known systematic factors. For instance, if we want analyse the calibration data of a balance, we know that systematically if we apply a mass of 1 kg we get an output very near to 1 kg, if we apply 2 kg again we get something very near to 2 kg, and so on. We have to report the data as they are, but we have to be aware that our interest is not the 1 kg or 2 kg nominal outputs, but on the differences of the outputs obtained from the nominal ones. Therefore, as first operation, we change the variables in such a way to get values that evidence more clearly possible effects.

b. application of statistical tests that allow to evidence the presence of systematic effects or accidents. We have different tests for systematic and accidents, therefore seems natural to ask if it is better to do one or the other before. It is better to clarify immediately that the tests shall be repeated, considering in any case a repetition from the beginning, any time the data set is modified because a systematic effect has been corrected or an outlier eliminated; therefore one test or the other before or after is not important: we apply before the test in the direction we guess the possible effects are bigger (if we see some strange data we apply before the detection of outliers, if it seems there is a possible tendency we before search for systematic effects). If test evidence an effect, the proper correcting action is done, the data corrected or excluded at the level of the initial data and all the analysis repeated. This because when a disturbance is eliminated, the tests become more sensitive and are able to eliminate a smaller disturbance they cannot see in presence of the bigger one;
c. for systematic effects, the previous step only evidences its presence. It is necessary to apply other test that indicate which are the factors producing the systematic effects;
d. again for systematic effects, when the factors which produce them have been detected, one must seek a mathematical model which describes the effect values, so that it can calculate the corresponding corrections.

The tests to be used to go this path are:

- for evidencing systematic effects any normality test. In fact the theorem of central limit says that if there are only multiple random effects, the distribution is normal. Therefore we adopt the null hypothesis that our data are normally distributed and apply any normality test to check it; if the test rejects the normality, there is evidence that something else than random effects (systematic or measurement accidents) is present. We will describe the Pearson's χ^2 and the Normal Probability Plot (NPP);
- for evidencing outliers exclusion principles. These statistical methods indicate if there are data with values that are unlikely as compared with the other data values, i.e. *outliers*. We will describe the Chauvenet method and IQR method;
- for connecting effects with possible factors, hypothesis test method and Analysis of Variance method, which will be described;
- for establishing a factor-effect relationship the regression method, which identifies a mathematical model.

The statistical methods to follow the described path will, therefore, be available. Remembering that test shall be repeated any time a modification is applied to the original data set, it is better to have a well-organized spreadsheet, to perform all the processing almost automatically.

6.3.1 The principles of exclusion

Taking experimental data, only in case an evident irregularity of the procedure is happened it is possible to discard a data taken. The result value, in principle, should not be considered at all. This shall be clear to the operator that must be careful not to succumb to the temptation to exclude all the data that are, in his opinion, too far away from the average.

This type of exclusion is the task of the principles of exclusion, which should be used with caution. In fact, the operation logic of the principles of exclusion has an important lack, which however is not unbridgeable with mathematical methods, whereby it needs to be offset with the attention and the experience. Their operating principle is related to the fact that accidents are rare events. A way for evidencing effects of measurement accidents is considering data which values that makes them commonly rare. The problem is that, while it is true that accidents are rare events, it is not true that all rare events are accidents. Let's talk about, for example, the metric instruments mentioned in section 3.2.1. They have a very low resolution. With these instruments it is very common to have a series of data that have all exactly the same value and only one or two are passed to the indication immediately subsequent or preceding. In this situation, a perfectly normal event (the indication to the next or previous digit for a random variation) is read as a rare occurrence.

The management of outliers is, therefore, questionable: certainly not just delete them in any case, but spend some attention for understanding if there is any real reason that justifies their presence. Some considerations on the risk of first kind involved, stated considering possible consequences of deletion, can help for taking a decision.

A very simple method was seen in section 5.4 is connected with the presentation of data using the box-plot. Connected with this method it is possible to evidence outliers in a really conventional way, connected with the *interquartile range* IQR, which is the difference between the third quartile Q3 and the first quartile Q1. The acceptance boundaries are set at Q1 - 1.5 IQR and Q3 + 1.5 IQR, but this convention penalise the evaluation with sets with large number of data. To correct this situation at Politecnico di Torino was proposed to use at the place of 1.5 the value given by the formula $1.5 \cdot \left[1 + 0.1 \cdot \log(n/10)\right]$.

In the case of normal distribution, it can be derived that the conventional *IQR* method assigns a fixed probability of 0.70% to have bilateral external data. Instead, in case of modified IQR method, that probability decreases with the sample size n (it is about 0.70% for n = 10, 0.45% for n = 50, 0.37% for n = 100).

Another method, called *Chauvenet's exclusion principle*, connects the probability of having an accident to the number of tests carried out. Chauvenet was an astronomer working in the U.S.A. in the middle of 18th century. His experience was to perform experiments in which *n* data where measured and he noticed that 50% of the experiments were completely correct and 50% had one or more accidents. Therefore he assigned the conventional probability of 50% to have an experiment with one outlier in its *n* data. This has a logical reason. More tests are made, more likely there are data belonging to the tails of the distribution, deviated from the expected value, which must not be considered measurement accidents.

Thereafter, it is easy to calculate with the binomial distribution the probability p to have an accident in one measurement when the probability of having 1 accident over *n* measurements is $\mathcal{P}(n,1)$. An approximated value for *n* large is $p = \mathcal{P}(n,1)/n = 1/(2n)$. Chauvenet principle considers data normally distributed as identified by *m* and *s* of the experimental data, divides p symmetrically between the lower tail and the upper tail and with that level of risk of first kind defines the boundaries of the tails. If one data falls within the tails, it is considered an outlier as described in Fig. 6.3.

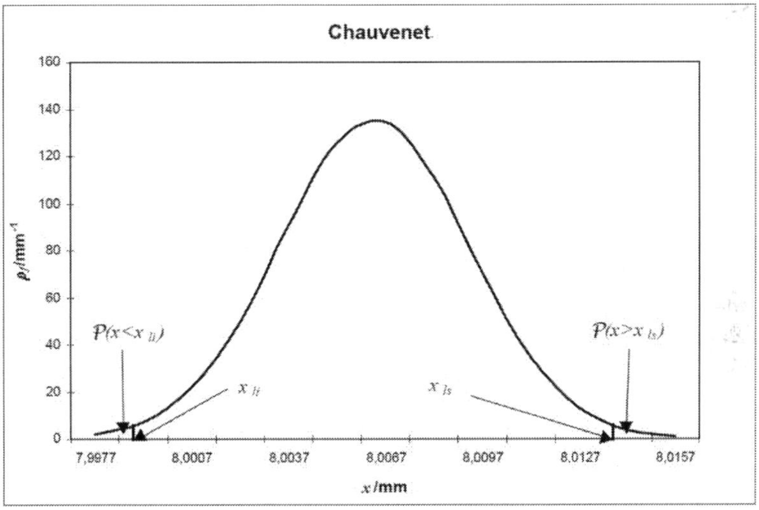

Fig. 6.3 Determination of the tail areas corresponding to the probability $P(\mathbf{x}<x_{li})$, that in one measurement the result will be lower than x_{li} (left tail), and to the probability $P(\mathbf{x}>x_{ls})$, that it will be bigger than x_{ls} (right tail). Given conventionally $P(\mathbf{x}<x_{li})$ e $P(\mathbf{x}>x_{ls})$, we can calculate the values of the boundaries of exclusion.

This convention produces variable results depending on the number of data, as shown in Table 6.6, along with the corresponding values of the standardized variable. The convention of the quartiles corresponds to the answers of more rigorous tests for a number of tests included between 50 and 100.

Applying the Chauvenet's convention to the data of Table 6.1, it is possible to easily calculate the cumulative probability to the upper limit and the values of z_{ls} and z_{li} corresponding to x_{ls} and x_{li}.

In the exemplified case n = 50 from which we easily obtain the values of $z_{ls}=2.58$ and $z_{li}=-2.58$, because of symmetry, and the values of the limits x_{ls} and x_{li} with the equation (5.42):

$$z_{li} \approx \frac{x_{li} - m}{s} \Rightarrow x_{li} \approx m + z_{li}s = 8{,}0063 - 2{,}58 \cdot 0{,}0029 = 7{,}9988$$

$$z_{ls} \approx \frac{x_{ls} - m}{s} \Rightarrow x_{ls} \approx m + z_{ls}s = 8{,}0063 + 2{,}58 \cdot 0{,}0029 = 8{,}0139$$

Comparing the values of the boundary so determined with the experimental data of Table 6.1, we can see that a value should be excluded: 8.0140. The table of the original data must remain unchanged, then a new table must be prepared (Table 6.7) with the correct data.

Tab. 6.6 Values of the bilateral probability and of the standardized variable to the lower limit z_{li} and the superior one z_{ls}

n	**Bilateral probability**	**Values of z_{li} and z_{ls}**
10	0.050	±1.96
20	0.025	±2.24
50	0.010	±2.58
100	0.005	±2.81

Tab. 6.7 Experimental data corrected after the detection of a probable measurement accident. The value of the box with the data highlighted in bold has been replaced with the average value obtained after excluding the original data.

Oper. 1	**Oper. 2**	**Oper. 3**	**Oper. 4**	**Oper. 5**
8.0038	8.0040	8.0060	8.0092	8.0053
8.0051	8.0025	8.0041	8.0104	8.0076
8.0037	8.0093	8.0028	8.0110	8.0113
8.0017	8.0025	8.0049	8.0087	8.0112
8.0023	8.0046	8.0025	8.0047	8.0066
8.0062	8.0020	8.0030	8.0076	8.0065
8.0080	8.0046	8.0073	8.0056	8.0114
8.0058	8.0085	8.0065	8.0113	8.0064
8.0031	8.0056	8.0071	8.0068	8.0065
8.0084	8.0037	8.0064	8.0093	8.0057

Having found an outlier, a common practice for deciding the real exclusion is as follows:

- repeat the measurement if possible and replace it with the new data. If the operator, correctly trained, has noticed the strange value, registered it thereafter repeated the measurement and given a replacement value in a note in case of exclusion, replace with the replacement data;
- if it is not possible to repeat the measurement and there is no replacement data, simply exclude the data;
- if your software does not work with a missing data, it is possible to adopt a trick (but, remember, it is a trick!). Cancel the outlier, do the average with the empty cell (this is usually allowed also with missing data) and fill the empty cell with the value of the average. This will not change significantly the average, but changes the variance and the distribution shape, therefore, whenever possible, use one of the first two alternative.

After the elimination, remember to repeat the test, because, cancelled a big mistake, smaller ones can become evident.

In our case we have replaced the excluded value with the average value obtained after the exclusion, in order to have again a complete table of results.

The replacement of the data excluded also leads to a variation of the average value (new value 8.0062), of the standard deviation (new value 0.0027), and of the table of the experimental frequencies, reported in Table 6.8.

Tab. 6.8 Distribution of the experimental frequency after the correction.

From [mm]	**To** [mm]	**Central value** [mm]	f_a	f_r	ρ_f [mm^{-1}]
7.99845	7.99995	7.9992	0	0.000	0.00
7.99995	8.00145	8.0007	0	0.000	0.00
8.00145	8.00295	8.0022	7	0.140	93.33
8.00295	8.00445	8.0037	7	0.140	93.33
8.00445	8.00595	8.0052	10	0.200	133.33
8.00595	8.00745	8.0067	11	0.220	146.67
8.00745	8.00895	8.0082	6	0.120	80.00
8.00895	8.01045	8.0097	4	0.080	53.33
8.01045	8.01195	8.0112	5	0.100	66.67
8.01195	8.01345	8.0127	0	0.000	0.00
8.01345	8.01495	8.0142	0	0.020	13.33

6.3.2 Normality test on the experimental distribution

If the variations of the results are due only to random effects, multiple and likely to produce both positive and negative variations, the expected experimental distribution is normal. Vice versa, if the experimental distribution is not normal, the presence of systematic factors can be assumed. These can be detected and corrected, so it is important to have the indication of their presence.

Examining the histogram of Fig. 6.2b, it is impossible to recognize if it has a normal distribution, not even overlapping it to our graphic. Random factors, in fact, can also strongly vary the shape of the histogram, which comes to assume a defined shape only when the number of data grows considerably.

To check the hypothesis of normality of the experimental distribution, we can use different methods. We will talk about two methods. The first, numeric and rigorous, is called χ^2 test and allows, if it is the case, to reject the hypothesis of normality at defined level of confidence $\mathcal{P}$, with a well-defined risk 1-$\mathcal{P}$ that the eventual rejection has no real reason.

The second one, graphic, is called Normal Probability Plot and does not allow to express the rejection of the hypothesis of normality with equal rigor, but, on the other hand, has the advantage of greater information content present in graphic form, which, with the diagram's trend, provides valuable information on the possible causes of non-normality.

6.3.2.1 Test of the χ^2

The χ^2 test was discussed in section 5.5.9: here it will be applied to the experimental data. We apply to the data of Table 6.7 as shown for the execution of the test of the χ^2, using the categories of Table 6.8. Some categories without data were inserted at the lower and upper end, to care to the tails of the compared normal distribution.

The theoretical frequencies f_{at} were obtained from the theoretical relative frequencies f_{rt}, themselves calculated as the theoretical probability in the same category intervals from the cumulative normal distribution.

For the data of Table 6.7 the results are shown in Table 6.9, using the standardized variable z in such a way that it is possible to make the calculation with the Table of the normal cumulative distribution $\mathrm{F}(z)$. In case of use of spreadsheet that allows the direct calculation of $\mathrm{F}(x)$, it is not necessary to pass through the variable z.

In the third and fourth columns are placed the values of the normalized variables z, calculated for the lower (z_i) and higher (z_s) category limits.

In the fifth column is placed the theoretical relative frequency, calculated as the difference of the cumulative distribution at the z_s and z_i category limits.

In the sixth column is calculated the theoretical absolute frequency as product of the relative frequency by the number of data (in the example, n=50).

The seventh column shows the experimental absolute frequency already determined previously (Table 6.1). Finally, in the eighth column are calculated the contributions of the different categories to W. The sum of the contributions gives the experimental value of W, which must be compared with the theoretical limits of χ^2 for the confidence level $\mathcal{P}$ required.

Tab. 6.9 Test of the χ^2

Category		z		Teor. rel. freq.			
From	*To*	z_i	z_s	$f_{rt} = F(z_s) - F(z_i)$	f_{at}	f_a	$(f_a - f_{at})^2 / f_{at}$
[mm]	[mm]						
7.99845	7.99995	-2,8341	-2,2843	0,0089	0,4	0	0,44
7.99995	8.00145	-2,2843	-1,7345	0,0302	1,5	0	1,51
8.00145	8.00295	-1,7345	-1,1847	0,0767	3,8	7	2,62
8.00295	8.00445	-1,1847	-0,6348	0,1447	7,2	7	0,01
8.00445	8.00595	-0,6348	-0,0850	0,2034	10,2	10	0,00
8.00595	8.00745	-0,0850	0,4648	0,2128	10,6	11	0,01
8.00745	8.00895	0,4648	1,0146	0,1659	8,3	6	0,63
8.00895	8.01045	1,0146	1,5644	0,0963	4,8	4	0,14
8.01045	8.01195	1,5644	2,1142	0,0416	2,1	5	4,10
8.01195	8.01345	2,1142	2,6640	0,0134	0,7	0	0,67
8.01345	8.01495	2,6640	3,2138	0,0032	0,2	0	0,16
						W	*10.29*

Let us examine the method of choice of $\mathcal{P}$, of which we can have an almost complete vision even without knowing the use that will be made of the measurement results. To choose the value of the confidence level it is necessary to evaluate the consequences of rejecting the null hypothesis. In our case, the statistical null hypothesis is that the experimental distribution is normal. This corresponds to the metrological hypothesis that the data are not affected by variable systematic errors or measurement accidents of such magnitude to significantly deform the distribution. If we accept a high risk of error to reject the hypothesis, we risk to have an indication of the presence of disturbing effects even if there are not. The consequence will be a greater focus in the analysis of the measurement method, analysis that must always be made in the common practice and for the rules of Quality. This means no additional costs; on the contrary, if our process passes a severe test (with high risk of refusal) one can avoid an internal inspection. We use, therefore, a risk of error of 15% that corresponds to a confidence level P =85%.

The shape of the theoretical distribution of χ^2 depends on the number of degrees of freedom, given by the number of possible differences (in this case corresponds with the number of categories examined, in fact for each category the absolute frequencies can be different from the theoretical absolute frequencies). From these differences we have to subtract the number of constraints (in this case in the comparison between normal distribution and histogram of the experimental distribution were imposed three constraints: equality of the average, of the standard deviation and of the total area). The degrees of freedom are, therefore, equal to the number of compared categories minus 3.

In the example done, that examines 11 categories, the degrees of freedom are 8, the confidence level P =0.85, therefore, the tables of the χ^2 or of the relative function in Excel calculate the minimum and maximum limits of the confidence interval:

Categories	11	Limits of the χ^2	
Freedom degrees χ	8	minimum	3.14
Confidence level P	0.85	maximum	14.27

The experimental value W can fall inside or outside the limits corresponding to a confidence level P. In the first case as in the example shown, the test does not provide information. The answer is that nothing unusual was detected, but there may be disturbances too small to be highlighted.

In case W is external, instead, the null hypothesis is rejected, at a confidence level P (there is only a probability 1 - P that the result is such by chance and not because of a real difference of the experimental distribution from the normal).

The test of the χ^2 is usually bilateral. The rejection is done both for values of W big (experimental distribution too different from the normal), both for values of W small (experimental distribution too similar to the normal). Since from the statistical point of view even this condition, which may seem desirable, is very improbable, therefore, is also object of suspicion.

6.3.2.2 Normal Probability Plot

The idea that underlies the method of the Normal Probability Plot (NPP) is the search for a presentation immediately recognizable of a form associated with the normal distribution.

The bell shape has no intrinsic characteristics that allow recognition. Even the shape of the cumulative normal distribution, an S (Fig 6.4), is not identifiable. In fact one of the few identifiable forms is the straight line, but the S shape deviates considerably from it.

To solve the problem of having a recognizable form, it has been prepared a suitable graphic, called *Normal Probability Plot* (*NPP*), with the ordinate scale deformed so that the S-curve of the cumulative normal distribution is represented by a straight line, as seen in Fig. 6.4.

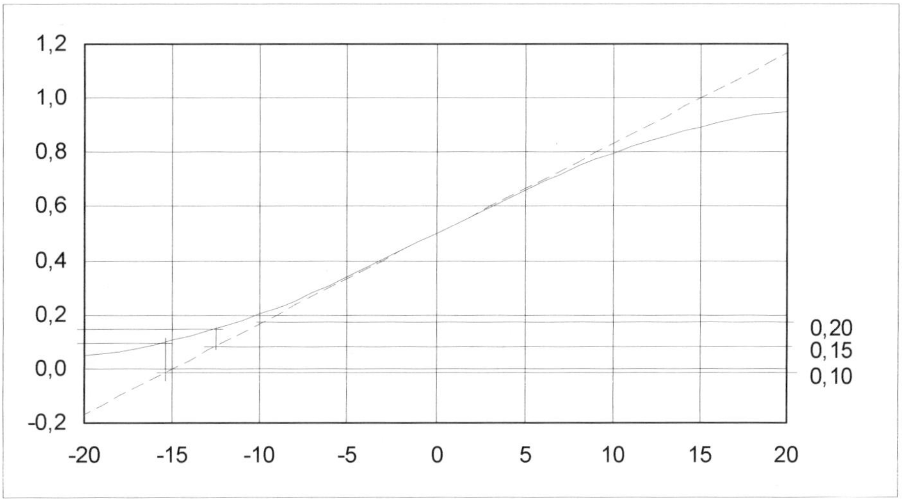

Fig. 6.4 *Example of how was generated the deformed scale of the NPP (right) using which the S shape of the normal cumulative distribution is represented by a straight line.*

Paper graphics with the ordinate so deformed are commonly marketed. Representing on them the trend of the experimental cumulative relative frequency it is possible to evaluate at a glance if the obtained trend approaches the rectilinear form, or deviates from it significantly. In the latter case, in addition to reject the hypothesis of normality of the experimental distribution, we can even get some information on the causes of non-normality.

The representation can be done using the n experimental data (i.e. table 6.10) that correspond to a normal distribution. In Table 6.10 data are reported in ascending order together with the ordinal number $\boldsymbol{i}$ and the corresponding relative cumulative frequencies. Recalling the definition of frequentistic probability, as the interval from -∞ to shortly before the first data do not contain any data, the associated cumulative probability is $f_{rc} = 0/n$. If we consider the interval from -∞ to just after the first data, it contains one data, therefore $f_{rc} = 1/n$. On the first data f_{rc} has a step from $0/n$ to $1/n$, therefore, if we want to define the value of f_{rc} corresponding to the first data, we shall use the central value of that interval, that is $f_{rc} = 0.5/n$. If we do the same for the interval from -∞ to shortly before and just after the second data, we can see that the step there is from $1/n$ to $2/n$, that is on the second data we shall use the value $f_{rc} = 1.5/n$. Check yourself if on the third the value is $f_{rc} = 2.5/n$. View that, it is easy to understand that on the ith data you can assign the cumulative relative frequency $f_{rc}=(i-0,5)/n$. This allows us to evaluate for each data the relevant relative cumulative frequency, as given in table 6.10. Having a NPP sheet, it would be possible to draw the graph. But it is even possible to use a worksheet to draw the graph, i.e. to deform the ordinate in the correct way. We can use equation 5.37 that show us a linear relationship between the standardized variable z and the physical variable $\boldsymbol{x}$ in case of normal distribution, therefore it is sufficient to evaluate from the values of cumulative probability the corresponding value of z

using the table or the function F(z) of the normal distribution to get the value of z corresponding to each value of x. Moreover at z=0 we have the average m of X, and at z=±1 there are differences ±s with respect to m.

Tab. 6.10 Table of normally distributed data, sorted in ascending order, with the corresponding cumulative relative frequency f_{rc}.

i	x/mm	f_{rc}	i	x/mm	f_{rc}	i	x/mm	f_{rc}	i	x/mm	f_{rc}
1	9.2973	0.01	14	9.2999	0.27	27	9.3009	0.53	40	9.3018	0.79
2	9.2977	0.03	15	9.3000	0.29	28	9.3009	0.55	41	9.3018	0.81
3	9.2985	0.05	16	9.3000	0.31	29	9.3010	0.57	42	9.3019	0.83
4	9.2986	0.07	17	9.3001	0.33	30	9.3011	0.59	43	9.3020	0.85
5	9.2988	0.09	18	9.3002	0.35	31	9.3012	0.61	44	9.3023	0.87
6	9.2988	0.11	19	9.3003	0.37	32	9.3012	0.63	45	9.3024	0.89
7	9.2990	0.13	20	9.3003	0.39	33	9.3016	0.65	46	9.3025	0.91
8	9.2991	0.15	21	9.3005	0.41	34	9.3016	0.67	47	9.3027	0.93
9	9.2994	0.17	22	9.3006	0.43	35	9.3017	0.69	48	9.3028	0.95
10	9.2995	0.19	23	9.3007	0.45	36	9.3017	0.71	49	9.3038	0.97
11	9.2996	0.21	24	9.3007	0.47	37	9.3017	0.73	50	9.3041	0.99
12	9.2997	0.23	25	9.3007	0.49	38	9.3017	0.75			
13	9.2997	0.25	26	9.3008	0.51	39	9.3017	0.77			

In Fig. 6.5 is shown the NPP of the Table 6.10 data. It can be observed that the trend of the experimental data is similar to a straight line[5]. In this case we cannot reject the hypothesis of normal distribution. Note, however, that the evaluation if the data are arranged properly on a rectilinear trend is arbitrary; it is not possible to assign to that a certain level of confidence. The disadvantage of the arbitrary rejection of the normal condition is compensated by an advantage deriving from the graphical description.

In case the distribution is *hypernormal* (bell curve higher at the centre than the normal distribution, and, being the total area of any statistical distribution always unitary, the extreme parts are lower) the graphic is presented as exemplified in Fig. 6.6. showing an S shape.

---

[5] We must consider random effects, which never allow a perfect alignment of the experimental points. The assessment can be done identifying the extent of localized disturbances, for example just before 9.302 in Fig. 6.5. These defects cannot be thought linked to a form error, therefore, are an index of the random effects.

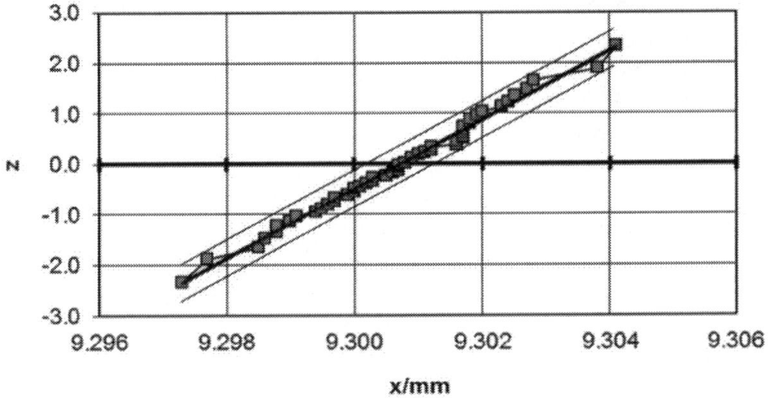

Fig. 6.5 *NPP for a normal distribution (data of Tab. 6.10). Errors and omissions lead to local irregularities compared to an exactly rectilinear trend. The condition of straightness must, therefore, be evaluated with a rectilinear band, whose width corresponds to the size of the localized defects, which must contain all the experimental points.*

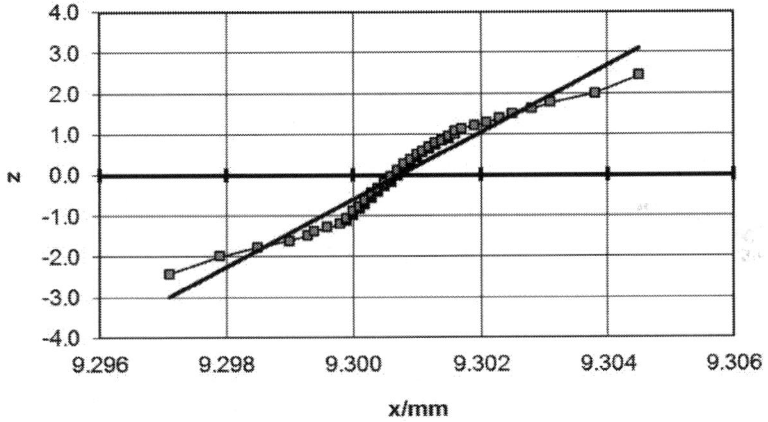

Fig. 6.6 *NPP for a hypernormal distribution*

From the experimental point of view this condition may be linked to a process of "filtering" that has eliminated preferentially data far from the average value. This situation is present both in the field of metrology and in the field of production. In the first case very often it takes a negative characteristic. The hypernormal distribution corresponds, in fact, to a reduction in the number of the data further away from the average value; this happens, in general, because the operator tends to exclude on the basis of personal judgment the more deviating data from the value that he expects. This practice should be discouraged, because it distorts the distribution of results and does not allow a correct calculation of uncertainty. The exclusion of results must be done only when it is recognized that the cause that has produced the deviation or using appropriate principles of exclusion as, for example, the principle of Chauvenet.

In case of production results, instead, the hypernormal indicates that the part of production examined was subjected to a selection process.

The opposite case is the *hyponormal* distribution, represented by a curve lower than normal in the central area and higher at the extremes, whose NPP is represented in Fig. 6.7. This condition may be due, generically, to the presence of phenomena that enhance the systematic differences from the average. Two typical effects are quite common, involving both the metrological and the production area: the presence of discontinuities or of tendencies. Discontinuity involves discontinuous factors; measurement made with few different instruments, production made with few different machines, systematic difference between operators, inherent discontinuity of the measurand. As an example of discontinuity consider, for instance, a mechanical balance with the pointer driven by a mechanism with a significant clearance. This means that the pointer is always half of the clearance lags behind the movement. As the balance reach the equilibrium point oscillating, sometime with a clockwise rotation, sometime counter- clockwise, the pointer has two different equilibrium positions, significantly separated. The corresponding distribution will have two values of maximum relative frequency, that is two modes and it is called bimodal. It is evident that this situation will produce a distribution that is larger, and being the total area equal to 1, lower, mainly in the centre, than the normal distribution.

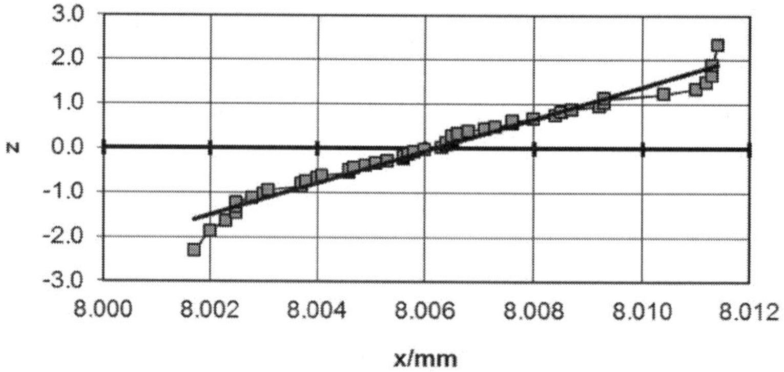

Fig. 6.7 NPP for a hyponormal distribution. Note the shape that has higher slopes at the extreme parts and lower slope at the centre. Being a representation of the cumulative probability, that is an integral of probability density, higher slopes mean that the density of probability at the extremes is higher than the normal distribution and lower at the centre.

Tendencies can act on measurement or production and are usually due to continuous factors: the natural change of temperature, wear conditions, the effect of non-linearity or hysteresis, etc. One continuous variable is usually forgot, but it is always present; the time, the sequence of the operations. In this case the measuring result belongs to a normal that moves with continuity for the tendency present. If this happens for a sufficiently long time, the global distribution, corresponding to the convolution of the various subsequent normal distributions, is significantly flattened in its central part. Tendencies can be seen very well in graphical form. For this reason specific software for statistical analysis automatically presents the graphs of the dependent variable versus each independent variable to show possible tendencies. For the time data are presented in their sequence, therefore it is important to input data in order of their temporal sequence. As said before, it is fundamental to clean the variables from the known systematic relationships; otherwise tendencies cannot be seen on the graphs. For instance common measuring instruments have frequently relative uncertainties at the level of 0.1%, i.e. the main errors are frequently smaller than 0.1% of the measured value. But it is impossible to see 0.1% on a graph (if you consider that it is well readable 1 mm, the smaller measurement value represented shall be on a scale 1 m long!), therefore the solution is to subtract from the output the nominal value to evidence the effect of errors. If you want to see the non-linearity, you have to take away the linear part.

Discontinuities and tendencies are represented by very similar hyponormal shapes of NPP. However, they are very different in substance. Therefore, the indication of hyponormality is only a clue that must be analysed to see which of the cases cited match the actual measurement or production conditions. In case of data reported in Tab. 6.1, the NPP is represented in Fig. 6.8.

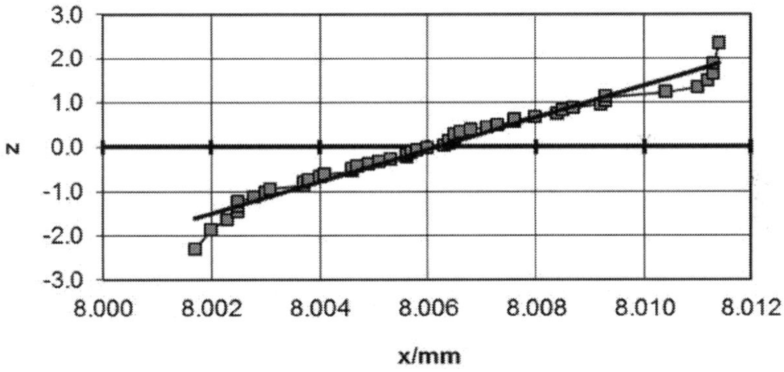

Fig. 6.8 NPP for the distribution of Table 6.7 data. The trend is in the form of a hyponormal.

It can be noted how the shape is of the hyponormal type. This requires to investigate the problem and to identify the presence of systematic factors, which were not reported by the method of χ^2, although we adopted for it a confidence level of 85%.

6.3.3 Comments on normality tests

As it turns out the two alternative evaluations of normality had a different sensitivity and thus give different answers. Note that even so, they are not incompatible. In fact, we emphasize here that the response of a statistical test can never be “the distribution examined is normal", but only either a rejection: "With a risk of error of 15%, the hypothesis of normal distribution is rejected", or a declaration of failure: "The test is not able to show significant discrepancies from the normal”. The χ^2 declared to be not able to see sufficient differences from the normal distribution to reject the null hypothesis, while NPP was able to see the differences. It is therefore clear that the two responses are not incompatible.

The NPP shows a clear hyponormal shape, therefore we have to consider that there are systematic effects and we shall start a research of the factors that produce systematic effects, using hypothesis tests and ANOVA.

7 IDENTIFICATION OF SYSTEMATIC FACTORS THE LINEAR REGRESSION

7.1 RESEARCH OF THE SYSTEMATIC FACTORS

The previously examined methods of the χ^2 and of the NPP, provide an indication of the presence of systematic factors and possibly on their type, but they do not manage to identify them. The research for the identification of systematic factors often requires an analysis of all the experimental process. However, when tendencies are involved, we can obtain a further indication examining the global evolution of the data. This is essential and often decisive. It reveals the internal relations that are not highlighted in a different way. For example, the trend of NPP of fig. 6.8 indicates the possible presence of a drift in the data reported in Table 6.7, that is a systematic tendency of the results obtained in subsequent times to vary in the same sense, then a variation that, apart from the irregularities caused by random factors, remains generally monotonic in function of time.

Not having registered the data time, we can, however, evaluate the presence of a drift reporting data in order of collection, as highlighted by the presentation of the data themselves in their temporal succession, represented in figure 7.1. From the trend of the data is possible to deduce the pattern of the systematic factor involved, which seems to have a linear trend versus the sequence of data.

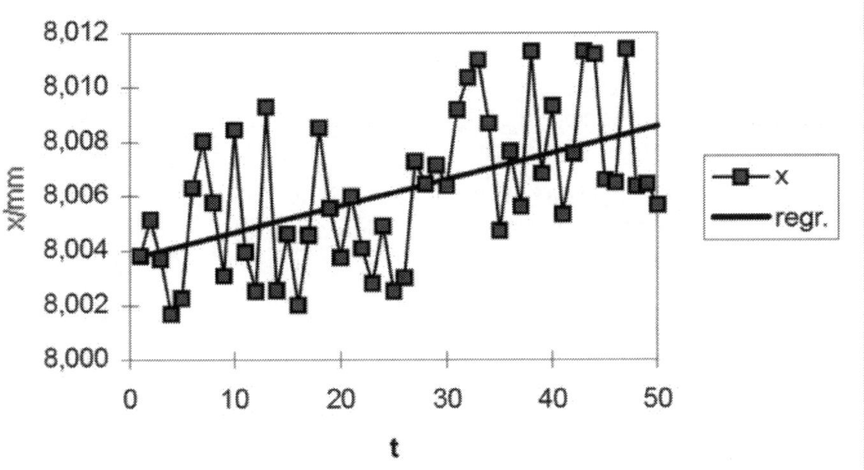

Fig. 7.1 Trend of the data of Table 7.1 represented in their temporal sequence. Notice how the graphical representation highlights an almost linear drift.

7.2 THE LINEAR REGRESSION

Regression methods[1] have wide use whenever we want to identify the functional relationship between one or more independent variables, often indicated with the symbol X, and the dependent variable, often indicated with the symbol Y. It is better to clarify immediately that there is nothing substantial in these symbols; they are used in most spreadsheet to indicate independent variables and dependent variable. Essential is to understand what are the independent variables, used as input data to do the calculations (frequently put in abscissa), and the dependent one, which is the result obtained (frequently put in ordinate). This fact is immediately seen in our case, because we want to identify the relationship of the drift, which is the variation of our measurement versus time. Therefore, we have to calculate the expected values of our results, x, taking the time, t, as input. The function we are seeking is in the form $x = f(t)$, so our independent variable is t and our dependent variable is x; but if we use for solving a spreadsheet, we will be asked on what is our x, and we shall put our values of t (note that it is not expressed the unit of measurement, since to see if there is a drift we have just used the sequence number), and what is our y, and we shall put our values of x. It is very useful, in order to overcome the mental constraint that connects the letter X to the abscissa and the Y to the ordinate, to describe a case in which the used variables have different names.

The steps can be summarized in three successive phases, which, if necessary, can be repeated:

- make an assumption of the mathematical model that represents the trend, taking into account the graphical presentation of the experimental data or considerations based on experience (technical standards, scientific literature);
- calculation of the parameters of the mathematical model with the *method of the Least Squares*;
- evaluation of the residuals, namely the differences between the experimental values and the values calculated with the mathematical model. This is used to decide whether the model used is adequate.

While the second phase is fully described by a mathematical process implemented in worksheets, the first and, in part, the last stage are entrusted to the experience of those who analyse the experimental data. A useful aid is obtained by using a graphical representation, which can indicate the shape of the trend and allows to guess which mathematical function can represent it: if a straight line will be sufficient or if more complex curves are necessary, such as parabolas, hyperbolas, logarithmic, exponential, sinusoidal or circumference[2]. It is good to use all the information on the phenomenon described by the

[1] To get a complete description see e.g. Draper N.R., Smith H., "Applied regression analysis", John Wiley, N.Y., 1966

[2] The term "linear" does not concern the trend of the function, but its coefficients to be calculated.

experimental data; so to choose, where possible, a mathematical model corresponding possibly to the physical laws involved.

The adopted mathematical model proposes only the trend line that describes the data. This consists of a generic equation that contains parameters to be determined to identify the curve of that family which best fits the experimental data. For example, if we decide that a straight line well represents our experimental data, we shall use the mathematical model $x = a_0 + a_1 t$. This will tell us that what is the dependent variable, x in the example, the independent variables, only t in the example, and the model parameters (or coefficients) a_0 and a_1. We emphasize again that it is important to distinguish which are the dependent and the independent variables; the assigned names and symbols have no importance.

In the second phase, parameters must be identified. The calculation of their value should be done with the Least Squares method, which can be easily applied if the mathematical function used is linear in the parameters themselves. But in some cases we shall use mathematical models which parameters are not linearly connected with the dependent variable. Two different cases are therefore to be considered, which go under the names of *linear regression* and *nonlinear regression.* The first is easy to calculate and is implemented in nearly all common spreadsheets; the second is more complicated because it requires a nonlinear calculation.

For instance, if we assumed that the trend of the experimental data can be well described by a parabola (nonlinear function considering the independent variables) which connects the dependent variable Y to the independent variable X by the formula:

$$Y = a_0 + a_1 X + a_2 X^2 \tag{7.1}$$

we see that, even if the type of the selected function is not linear, the Y is linear as a function of the parameters a_0, a_1 and a_2 that have to be determined to extract from the family of parabolas thus described the one that fits the experimental data.

Whenever this condition of linearity occurs, we can use the method of the *linear regression*.

The same rule applies, for example, if we use an exponential function like:

$$Y = a_0 + a_1 \cdot e^X \tag{7.2}$$

because, again, it is linear in the parameters.

If we want, instead, to use an exponential function like:

$$Y = a_0 + e^{a_1 X} \tag{7.3}$$

this would be a non-linear regression, much more complex.

In the third step we must validate the guessed mathematical model: assess whether the selected model is suitable for the experimental data. This evaluation is carried out taking into account the purpose for which we search for a mathematical model: identify systematic factors and their trends. If the chosen mathematical model corresponds to the systematic factor present, then the *residuals* (difference between the experimental data and the corresponding ones calculated with the mathematical model) are due only to random factors, so they should have a normal distribution. If this hypothesis is not excluded by a normality test (usually the NPP), then we can accept the chosen model.

7.2.1 Calculation of the linear regression

All linear forms described above can be represented by a function with multiple variables:

$$Y = G\left(X_1, \ldots, X_j, \ldots, X_q\right) \tag{7.4}$$

in which the independent variables X_j are the various contributions of the selected function. For instance, for the parabola is used:

$$\begin{aligned} &Y = a_0 + a_1 X + a_2 X^2 \qquad X_1 = X \qquad X_2 = X^2 \\ &G(X_1, X_2) = a_0 + a_1 X_1 + a_2 X_2 \end{aligned} \tag{7.5a}$$

and for the exponential described by the equation 7.2 is used:

$$\begin{aligned} &Y = a_0 + a_1 e^X \qquad X_1 = e^X \\ &G(X_1) = a_0 + a_1 X_1 \end{aligned} \tag{7.5b}$$

The variables X_j can also be completely independent factors. For example, X_1 may be the time, to see if the temporal succession influence the results, as in our example, X_2 may be the ambient temperature at which the measurement was taken, X_3 atmospheric pressure, and so on. In this case, we often talk about multiple regression.

We can switch from the relationship between variables to the one between the experimental data:

$$y_i = G\left(x_{i1}, \ldots, x_{ij}, \ldots, x_{iq}\right) + \varepsilon_i \tag{7.6}$$

in which the experimental error ε_i is totally attributed to the dependent variable Y. This is a rigorous condition in a strictly mathematical point of view. From the practical point of view, the error is due in part to the experimental errors on the data x_{ij}, mostly to the errors on the corresponding values y_i of the dependent variable Y and also to the inadequacy of the mathematical model (remember chapter 1, correlation uncertainty).

Only when the contribution due to the uncertainty on the dependent variable Y is predominant, the usual linear regression can be applied in a strict manner. However it should be noted that even when we depart, not too much, from this condition, results that well indicate the behaviour of the variable Y are obtained. Therefore, this method is often used to obtain a statistical interpolation of experimental data that, while not containing rigorously information of the regression variance, is always a great help to understand the functional relationships of the observed phenomenon.

7.2.2 *Method of the least squares*

To determine the values of the coefficients, we carry the examination on a concrete situation in which there is a functional relationship, as the one described by the equation (7.4) and, therefore, a relationship between the data represented by (7.6), which can always be linearized as described by the examples of (7.5a) and (7.5b).

The linearized equation must apply to all the experimental points. So we can write a system of n equations in which the experimental values x_{ij} and y_j are known, while the parameters a_j are unknown. As we are aware of random errors, this system of equations shall not be deterministic in the sense that it shall have more equations than unknown variables. The number n of tests shall be larger than the number q of the model parameters. Therefore, it should be solved with the *method of the least squares*. Let's write the system in its general form

$$\begin{cases} y_1 = a_0 + a_1 x_{11} + \ldots + a_j x_{1j} + \ldots + a_q x_{1q} + \varepsilon_1 \\ \cdot \\ y_i = a_0 + a_1 x_{i1} + \ldots + a_j x_{ij} + \ldots + a_q x_{iq} + \varepsilon_2 \\ \cdot \\ y_n = a_0 + a_1 x_{n1} + \ldots + a_j x_{nj} + \ldots + a_q x_{nq} + \varepsilon_n \end{cases} \tag{7.7}$$

that can be expressed in matrix form:

$$\begin{Bmatrix} y_1 \\ \cdot \\ y_i \\ \cdot \\ y_n \end{Bmatrix} = \begin{bmatrix} 1 & x_{11} & \ldots & x_{1j} & \ldots & x_{1q} \\ \ldots & \ldots & \ldots & \ldots & \ldots & \ldots \\ 1 & x_{i1} & \ldots & x_{ij} & \ldots & x_{iq} \\ \ldots & \ldots & \ldots & \ldots & \ldots & \ldots \\ 1 & x_{n1} & \ldots & x_{nj} & \ldots & x_{nq} \end{bmatrix} \begin{Bmatrix} a_0 \\ a_1 \\ \ldots \\ a_j \\ \ldots \\ a_q \end{Bmatrix} + \begin{Bmatrix} \varepsilon_1 \\ \cdot \\ \varepsilon_i \\ \cdot \\ \varepsilon_n \end{Bmatrix} \tag{7.8}$$

that is:

$$\{y\} = [1, x]\{a\} + \{\varepsilon\} \tag{7.9}$$

We need to determine the matrix $\{a\}$ which solves the system. In order to do this, we pre-multiply both sides of the equation for the transpose of $[1, x]$:

$$[1, x]^T \{y\} = [1, x]^T [1, x]\{a\} + [1, x]^T \{\varepsilon\} \tag{7.10}$$

The solution can be formally obtained in a very simple way if the matrix $[1, x]^T \{\varepsilon\}$ is null. Let's see what is the shape of such matrix:

$$\begin{bmatrix} 1 & \dots & 1 & \dots & 1 \\ \dots & \dots & \dots & \dots & \dots \\ x_{1j} & \dots & x_{ij} & \dots & x_{nj} \\ \dots & \dots & \dots & \dots & \dots \\ x_{1q} & \dots & x_{iq} & \dots & x_{nq} \end{bmatrix} \begin{Bmatrix} \varepsilon_1 \\ \cdot \\ \varepsilon_i \\ \cdot \\ \varepsilon_n \end{Bmatrix} = \begin{Bmatrix} \sum_{i=1}^{n} \varepsilon_i \\ \cdot \\ \sum_{i=1}^{n} \varepsilon_i x_{ij} \\ \cdot \\ \sum_{i=1}^{n} \varepsilon_i x_{iq} \end{Bmatrix}$$

Note that this is a column matrix in which the first element is given by the sum of ε, the second by the sum of the products of ε for the values assumed by each of the variables.

In analogy to what happens for the average, for which we have seen that the sum of the residuals is null, we demonstrate[3] that the sum of the values assumed by a random variable ε and the sum of the products of random variables ε and x are zero.

In that case we have:

$$[1,x]^T\{y\} = [1,x]^T[1,x]\{a\} \tag{7.11}$$

---

[3] These conditions correspond to the choice of the values of the parameters $\{a\}$, so that the sum of the square of residuals $\{\varepsilon\}$ between the experimental values $\{y\}$ and the regression curve ones $[1,x]\{a\}$ is minimum. In fact the sum of the squared deviations is given by:

$$\sum_{i=1}^{n} \varepsilon_i^2 = \{\varepsilon\}^T\{\varepsilon\} = \left[\{y\} - [1,x]\{a\}\right]^T\left[\{y\} - [1,x]\{a\}\right] =$$

$$= \{y\}^T\{y\} - \{y\}^T[1,x]\{a\} - \left[[1,x]\{a\}\right]^T\{y\} + \left[[1,x]\{a\}\right]^T[1,x]\{a\} =$$

$$= \{y\}^T\{y\} - \{y\}^T[1,x]\{a\} - \{a\}^T[1,x]^T\{y\} + \{a\}^T[1,x]^T[1,x]\{a\}$$

The condition of minimum is given by the system of equations involving the partial derivatives:

$$\left\{ \frac{\partial \sum_{i=1}^{n} \varepsilon_i^2}{\partial a_j} \right\} = -2[1,x]^T\{y\} + 2[1,x]^T[1,x]\{a\} = -[1,x]^T\{y\} + [1,x]^T[1,x]\{a\} = 0$$

And substituting in (7.10):

$$[1,x]^T\{\varepsilon\} = [1,x]^T\{y\} - [1,x]^T[1,x]\{a\} = 0$$

that can be solved pre-multiplying for the inverse of $[1,x]^T[1,x]$:

$$\left[[1,x]^T[1,x]\right]^{-1}[1,x]^T\{y\}=\left[[1,x]^T[1,x]\right]^{-1}[1,x]^T[1,x]\{a\}$$
$$\{a\}=\left[[1,x]^T[1,x]\right]^{-1}[1,x]^T\{y\} \tag{7.12}$$

The resulting regression curve is the one that minimizes the sum of the squares of the deviations ε_i between experimental values y_i of the dependent variable and the corresponding values of the regression curve. The values of the parameters a_i can be calculated using the formula given above, or by using the linear regression functions implemented in spreadsheets.

7.2.2.1 Example of application on the experimental data

In case of the experimental data represented in Figure 7.1, it can be observed that the trend can be guessed linear. It is then used the model:

$$x = a_0 + a_1 t$$

The experimental data represent the x in function of t; therefore, in this case, x is the dependent variable and t the independent one.

The calculation of the coefficients a_j provides values a_0 =8.00370 and a_1 =0.00010, which lead to the regression line shown in Figure 7.1.

Not always the experimental data show so evident residuals compared to the regression line. In order to evaluate the model, it is always appropriate to represent the residuals respect the regression, to make sure to be able to use a suitable scale that makes them evident. This has been done in figure 7.2.

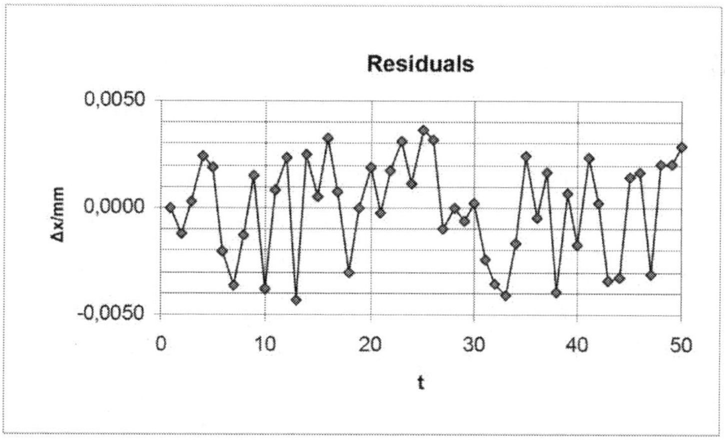

Fig. 7.2 Graphic of the residuals with respect to the regression line. These are the deviations between the experimental values of x and the values predicted by the regression equation.

The trend of the residuals is a useful indicator of the goodness of the model adopted. In fact, if there is a systematic factor, and that factor is identified and corrected, the residuals maintain the present random effects, arranging themselves with a random pattern. If, instead, the systematic factor has not been fully corrected, its presence is denounced by a certain content of "regularity" in the pattern of the residuals. A simple and efficient method to evaluate this "regularity" is to look at the sequence of the signs of the residuals. If there are only random effects, it is casual to have positive or negative residuals, so there is a frequent alternation of signs; if, instead, there is a systematic factor, the signs tend to group together.

In the examined case we can see in figure 7.2 that the alternation of the signs is good. Therefore the linear model can be considered adequate. For a more rigorous evaluation of the goodness of the chosen model, we can apply a normality test (χ^2 or NPP), as discussed in section 6.2.2.

To show a case in which the model of first order does not fully describe the systematic contribution, we examine the characteristic of a dynamometer: the data are reported in Table 7.1, with the regression residuals obtained with functions of first and second order.

Examining the signs of the first order residuals (Residuals 1), it can be seen how their signs are strongly grouped; therefore, the first order model is not appropriate. In the case of use of second order regression, instead, the residuals show adequate alternation of the signs. The graphical representation of Figure 7.3 shows this situation.

Tab. 7.1 Experimental data on the dynamometer characteristic. Results of the regression with first and second order functions.

Force [N]	Output [mV/V]	Regression [mV/V]	Residuals 1 [mV/V]	Regression [mV/V]	Residuals 2 [mV/V]
0	0.00000	0.00023	-2.3E-04	-0.00004	3.8E-05
100	0.19998	0.20019	-2.1E-04	0.20008	-1.0E-04
200	0.40014	0.40015	-1.2E-05	0.40017	-3.0E-05
300	0.60035	0.60011	2.4E-04	0.60022	1.3E-04
400	0.80020	0.80007	1.3E-04	0.80023	-3.4E-05
500	1.00030	1.00003	2.7E-04	1.00021	8.8E-05
600	1.20020	1.19999	2.1E-04	1.20015	4.7E-05
700	1.39990	1.39995	-5.0E-05	1.40006	-1.6E-04
800	1.59988	1.59991	-3.0E-05	1.59993	-4.8E-05
900	1.79979	1.79987	-7.9E-05	1.79976	2.9E-05
1000	1.99960	1.99983	-2.3E-04	1.99956	4.2E-05

Increasing the order of the function (i.e. increasing the number of independent variables) the number of degrees of freedom decreases. We can arrive at a function that passes exactly on all the experimental data, but this has no use. The aim is to separate the systematic part from the random one, in order to be able to treat the two parts in an appropriate manner. The artificial elimination of the random part, instead, prevents us to know the variability field of the data and, therefore, to correctly evaluate the uncertainty associated.

One way to proceed, frequently adopted for its simplicity, consists in the selection of a polynomial function as a template and to increase the order to obtain a better accordance with the experimental trend, as seen in the example of the characteristic of a dynamometer. On this approach we must, however, emphasize two warnings:

- we must take into account the physical behaviour of what we want to describe. If, for example, there are reasons to believe that the phenomenon is asymptotic, it can be described by a hyperbola, or approximated by a constant minus a negative exponential, but not by a polynomial form. The adopted mathematical model has to be suitable for the phenomenon that we want to describe. The polynomial model is valid in a broad way for those phenomena which can be described with a suitable development in a power series.
- the order of the polynomial used as the model should be increased only until obtaining a residual behaviour free of "regularity" suggesting the presence of systematic factors. An excessive increase in the order of the polynomial leads to an artificial tracking of the random factors and to a model identification that highly changes with the increase or replacement of even a few experimental data.

These observations emphasize the real difficulty of the application of the method. These are connected to what we said from the beginning: the difficulties of choosing the mathematical model to be adopted and subsequent assessment of its goodness.

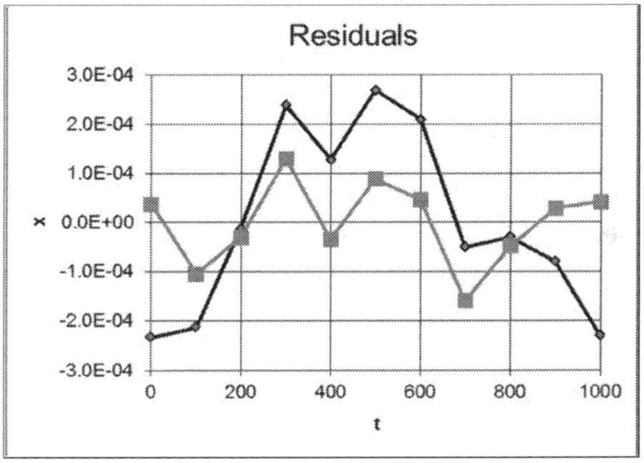

Fig. 7.3 Trend of the residuals with respect to the polynomial regression lines of first and second order. Note how, in case of first order, the non-linearity is highlighted as systematic factor, corrected by the second order regression.

We can use the same experimental data to evaluate what has been said in section 7.2.1 about the opportunity to choose, as independent variable, the one that has lower uncertainty. In case of measuring instruments, this request (still more rigorous from the point of view of mathematics, which assumes a null variance for the independent variables) is often disregarded for practical reasons. In a calibration, as the one described in Table 7.1, the known variable with the lower uncertainty is the one that describes the values of the reference standards. In the example, these are the forces applied by the force standard machine. Assuming this independent variable and the output as the dependent variable, a relation with which, known the force, we can calculate the output corrected from the systematic effects, is found as regression equation (in the case presented it can be clearly seen the linear error correction). In normal use we make the measurement of an unknown force, then comes to be known the output of the instrument and, from this, we want to calculate the value of the measured force; therefore we would prefer to have a mathematical model giving the force as dependent variable versus the output as independent variable. This situation is quite general for each instrument: in normal use the value of the measurand is not known, while with the measurement operation becomes known the instrument output. We want to have an equation to calculate the value of the measurand from the detected output value. To distinguish between these two situations, we will call the first "Equation of calibration" and the second "Equation of use."

From a practical point of view the results obtained with these two equations are, for instruments with regular metrological characteristics, fully compatible; therefore it is possible to use the most comfortable equations of use without problems. In many cases, the certificates of the Calibration Laboratories contain both the calibration equations and the use equations.

7.2.2.2 Subsequent evaluation of the model goodness

To evaluate the goodness of the mathematical model adopted is almost essential to examine the residual graphic, which globally presents the situation. It can be helpful in the evaluation the use of the means that enable to assess whether the assumption of normal distribution for the residuals should be rejected, suggesting that systematic factors are still present.

The residuals with respect to the regression can be treated as any random variable, and analysed, as it has been seen previously, with the test of the χ^2 or with the Normal Probability Plot. Let's examine the NPP obtained from the data of Table 7.1, respectively represented for the residuals of the regression function of the first and second order in Fig. 7.4 and Fig. 7.5.

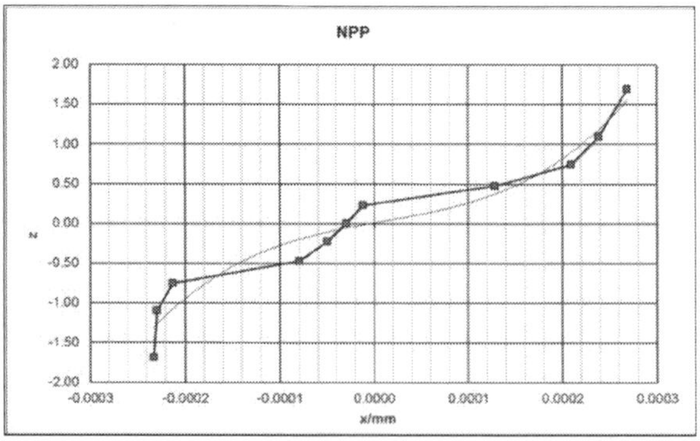

Fig. 7.4 Normal probability plot of the residuals with respect to the first order regression function. Note that the trend indicates a hyponormal distribution, as evidenced by a cubic trend-line.

The examination of these figures requires some experience. At first glance, in fact, the two trends seem equally deviate from the rectilinear form. We must, however, observe that in the first case the curved trend is well defined (we come to this conclusion if we try to represent the form with a band that has the width specified by the terms of irregularities more evident: the first two data after the zero). Even drawing a rectilinear portion with that width is not possible to contain all data.

In the second case, instead, local irregularities are significantly more (note the sequence of the five points just above zero); therefore, all the points are largely compatible.

It is also known how the NPP, for the case represented in figure 7.4, does not give a more evident indication than what shown by the residuals trend of Figure 7.3. This underlines the importance of examining the residual graphic.

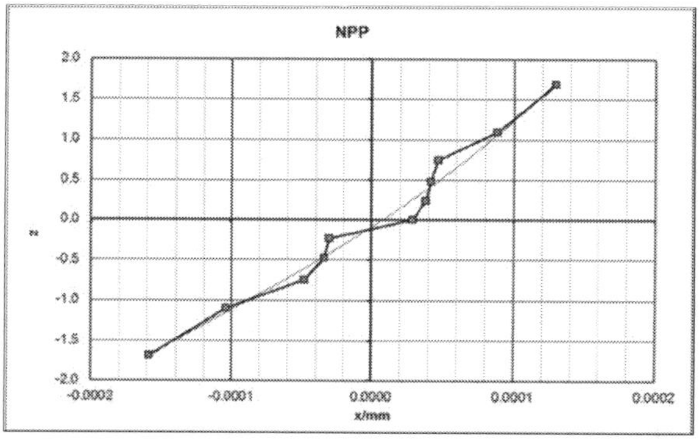

Fig. 7.5 *Normal probability plot of the residuals with respect to the second order regression function. Note that the trend, even with its irregularities, is not significantly curved, as shown by the cubic trend-line that is nearly rectilinear.*

8 HYPOTHESIS TEST AND INTRODUCTION TO THE VARIANCE ANALYSIS

8.1 *INTRODUCTION*

The importance of identifying the presence of systematic factors in the distribution of experimental data has been frequently underlined. Normality tests give the warning, but it is necessary to try to establish a connection to the factors that can cause systematic effects. A possible way is to put under control the factors suspected for evaluating whether their influence on the results is systematic or just random. For example, we might wonder whether the difference on production batches made by different workshops is significant, to highlight the possible systematic influence of the different operating modalities. Or we can be interested in understanding if the difference between the results obtained by some measuring devices in different environmental conditions is systematic, and if that is caused by the instruments or if it is due to the changes in environmental conditions.

It is clear that the decision on correcting actions must be linked to a concrete evidence of the significance of the causes, because it would be highly inefficient to investigate differences due only to random effects.

It would be equally inefficient to investigate all possible influence factors; therefore, this investigation activity requires two types of skills: one specific for the technological sector that we are now considering (manufacturing or measuring) and another one on the general analysis of the experimental results. For the first an expert of the specific production examined or of the specific measurement method is required, which has developed, with the experience, a sensitivity to identify possible functional relations that produce systematic effects. For the second, we need to apply appropriate statistical methods that will be described here.

In Chapter 6, a method for detecting the presence of systematic factors, based on the normality test, χ^2 and the normal probability plot, has been described; it can detect the presence of systematic factors when their effect produces a significant distortion of the experimental data distribution. Thereafter, for understanding which factors are guilty, statistical methods as hypothesis tests and analysis of variance.

The use of the hypothesis test

The problem is faced dividing the experimental results in groups, homogeneous for the presence of the factor that we want to analyse (factor under control). For example, if we want to detect if there are systematic differences between operators, we will split the data into groups produced by each operator. In this way we can apply statistical tests based on the assumption that there are no systematic differences between the groups and see if the test rejects this hypothesis. In fact, in the latter case, it is possible to improve the process identifying systematic effects which, thus, might then be corrected, compensated or eliminated.

If the analysis of the experimental data of table 6.7 (recalled here in Table 8.1 with, in addition, the average for each operator) will have indicated the presence of systematic effects, one suspect can be on systematic differences between five different operators. Then we have to see if the behaviour of the operators is homogeneous or are there factors that introduce systematic differences.

Groups can also intersect. If the operators have been asked to make measurements at specific times of the day, reported in the first column, you could also evaluate whether there is a systematic effect related to the time at which the measurement was performed.

There may also be more complex situations. Let's consider an industrial production made in two factories, three departments in each factory, four machine tools in each department, three operators for machine tool. To try to improve the production, it is appropriate to ask whether there are systematic effects and what or who has caused them. Factories, departments, machine tools or operators? It is obvious that the corrective actions are different for each case. It is also evident the great advantage of a consolidated indication on the main causes of variability.

Tab. 8.1 Results, expressed in millimetres, of the length measurements of a workpiece, obtained by five different operators. Single factor scheme of control.

Time	Oper. 1	Oper. 2	Oper. 3	Oper. 4	Oper. 5	Grand Average
8	8,0038	8,0040	8,0060	8,0092	8,0053	
9	8,0051	8,0025	8,0041	8,0104	8,0076	
10	8,0037	8,0093	8,0028	8,0110	8,0113	
11	8,0017	8,0025	8,0049	8,0087	8,0112	
12	8,0023	8,0046	8,0025	8,0047	8,0066	
13	8,0062	8,0020	8,0030	8,0076	8,0065	
14	8,0080	8,0046	8,0073	8,0056	8,0114	
15	8,0058	8,0085	8,0065	8,0113	8,0064	
16	8,0031	8,0056	8,0071	8,0068	8,0065	
17	8,0084	8,0037	8,0064	8,0093	8,0057	
Average	8,0048	8,0047	8,0051	8,0085	8,0079	8,0062

Groups are usually made on the basis of objective conditions: different operators, different machine tools or different working conditions. It should be evidenced that sometimes it may be useful to group them according to experimental evidence of possible systematic differences. For instance, in our example the box-plot of the data of Table 8.1 (see Figure 8.1) shows two groups that could be systematically different: the first of operators 1, 2 and 3, the second of operators 4 and 5.

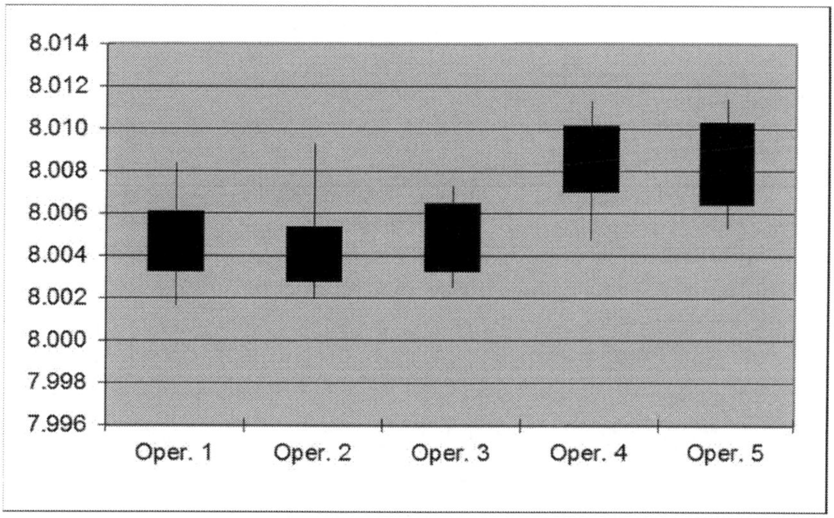

Fig. 8.1 *Box-plot that shows with the limits of Q1 and Q3 possible systematic differences between the group of operators 1, 2 and 3 and the second group of 4 and 5.*

Statistical methods of the hypothesis test or of the Analysis of Variance (ANOVA) can be applied to data grouped in that way.

8.1.1 Hypothesis test

The method of the hypothesis test allows to examine a single factor. It consists in formulating a hypothesis related to the behaviour of the factor that you want to analyse. The null hypothesis that is formulated will never be accepted, but only rejected. This condition imposes to draw the starting hypothesis, called the null hypothesis, in the opposite direction to the presence of the factor to be controlled. The hypothesis of significance of the examined factor will never be declared acceptable, but the hypothesis that it is not significant can be rejected, at the chosen risk of error.

To reach this answer, usually, we have to follow some steps:

a) choose how to represent numbers and locate in groups the data representative of the influencing factor hypothesized;
b) choose the statistical distribution which is believed to represent the analysed phenomenon;
c) determine the risk of error α of first kind that we are willing to take;
d) evaluate the boundaries of acceptance or rejection zones of the hypothesis. It is an operation that, for known distributions, can be easily done at the required confidence level, as seen at sections 5.1.6 and 5.1.8.

In evaluating the possible differences of result between the operators, we can assume the null hypothesis that there is no significant effect.

The points a), b), c) and d) previously seen are, in this case, defined as follows:

a) the behaviour of each operator is described by the average value of the n results he produced;
b) the question about the presence of a systematic factor can be asked from the statistical point of view in several ways: for example:
   1. is there homogeneity between the results produced by different operators? (Or equivalent: the averages of all the operators are pertaining to the relevant distribution of the averages?);
   2. do the differences between two average values obtained by any operator unaffected by systematic effects? (Or equivalent: the differences of the averages of all the operators are pertaining to the relevant distribution of the difference of averages?)
c) determine the level of confidence based on the effects of possible risk of errors[1]. In particular α error of the first type;
d) evaluate, on the basis of the risk of error α, the confidence limits of the adopted distribution.

Let us examine the first case in which the operators are evaluated from the average values of their measurements. As said before, the reference distribution is the distribution of the averages of q results; therefore a normal distribution with the grand average as average. About the standard deviation we shall do some observations: it is certainly possible, having five average values obtained from the five operators, to estimate from them the variance of the distribution of averages, but that assessment is poor because it is based only on four D.o.F. More convenient is to use the standard deviation s of all data[2] (49 D.o.F.) and determine, from it, the standard deviation of the average s_m in a more robust way:

$$s_m = \frac{s}{\sqrt{q}} \tag{8.1}$$

[1] In this case it is not appropriate to use the conventional 95% confidence level. In fact, the consequences of the risk of error α of the first kind are to spend some time doing the performance evaluation of the measurement procedures, only because the test showed an irregularity that does not exist. It is always advisable, when establishing the confidence level, to keep in mind the risk of error β of the second type: in this case, to be not warned on an irregularity even if it is present. A confidence level of 80%, i.e. the risk to one uselessly procedure revision over five, may be suitable for the present case.
[2] This method, although consolidated, leading to take as an indication of the natural variability even the effect of the differences due to the factor under control (in the case of the operators). Therefore, it decreases in some entity the sensitivity of the test, which remains well valid if only one, or few, outward expressions of the factor under control (in the example only one or a few operators) behaves in a different way. In case the differences were widespread, they would be taken as natural variability. The analysis of variance exceeds this limit with a more targeted evaluation than the natural variability.

Established the acceptable risk of error α, we can then estimate the limits of the confidence interval calculating its extremes with the statistical functions of spreadsheets or with tables: the z_s value of the upper limit corresponding to a cumulative probability $1-\alpha/2$. The limits result to be $m-z_s s_m$ and $m+z_s s_m$. The values calculated as described above are given in Table 8.2, along with the average values obtained by different operators. From the examination of the table, we see that the data corresponding to the first, second, fourth and fifth operator fall outside the confidence limits.

Tab. 8.2 Determination of the confidence interval limits and of the average values obtained by different operators

Parameters of the distribution of averages		Operator	Average
Number of data	10	1	**8,0048**
Average m	8,0062	2	**8,0047**
Standard deviation s_m	0,0009	3	8,0051
Confidence level $\mathcal{P}$	80%	4	**8,0085**
Lower limit	8,0051	5	**8,0079**
Upper limit	8,0073		

From this analysis we can draw our conclusion as follows: with the risk of error of 20% we reject the null hypothesis that the operators are homogeneous; therefore we can affirm that the differences between their mean results are systematic.

For the second question we compare all possible differences between the operators to the confidence interval of the relevant distribution, which is a normal distribution with zero as average (the null hypothesis says that the differences are not systematic) and as standard deviation the standard deviation of the difference of two mean values, which is the sum of the two variances, equal because pertaining to the same distribution. We obtain, therefore:

$$d = m_1 - m_2$$

$$s_d = \sqrt{\sigma_{m_1}^2 + \sigma_{m_2}^2} \approx \sqrt{s_m^2 + s_m^2} = s\sqrt{\frac{2}{q}} \qquad (8.2)$$

It is important to underline a logical bug, even if it is normally accepted in the practical application of the hypothesis test. We can observe that the control value, the confidence interval, is calculated from the standard deviation of the individual data. But, if the suspected systematic effects are present, the standard deviation increases and the confidence interval are enlarged. Therefore it is possible that, due to the presence of systematic effects, the test could be not able to discover it.

There is also a more dangerous and more complex drawback, not specific of the hypothesis test, dangerous for any experimental activity, well evidenced by the examined situation. This is very useful for teaching purposes, because it enables to show that is always necessary to criticise the data used.

Sometimes the data do not represent what they are supposed to. A situation is subtle when on the data act simultaneously two or more factors, in such a way that it is not possible to distinguish whether the effect is to be attributed to one or to the other.

These data have been presented as an example, to emphasize the importance of the sampling, which is the operation that provides the data to operate on. From the previous analysis in section 7.7.2.1 it was identified the presence of a drift which can be superimposed to the effect of operators: the drift acts on the sequence data and the operators have taken the data in succession (from the first operator who did measurements when the drift effect was low, to the last operator which data were affected by drift at its higher level). We cannot know whether, for example, the differences between the data of the first and last operator are due to their systematic behaviour or to the drift. Problems of this type are important and cannot be solved after the data have been taken. We must act before, in the organization of the experiment with specific techniques[3].

8.2 ANALYSIS OF VARIANCE

The analysis of variance (ANOVA) has a minimum disadvantage, as it indicates only the factor that produces the systematic effect and not the element of this factor that might be guilty. On the other hand, it has at least two great advantages: it can simultaneously handle more than one factor, operating even on complex schemes, and it avoids to use as control a value which can be influenced by the presence of systematic factors that are being sought.

The ANOVA test is based on the relation of two different calculations of the experimental data variance, so that one does not contain the effect of any systematic factor that we want to control, while the other is very sensitive to the presence of systematic effects on controlled factors.

As a first step, we must identify the factors that we want to check for the presence of systematic effects. On the basis of Table 8.1, we consider a division into groups for the factor "Operator," which may give different results on different columns of the data table, and the factor "hour of measurement", which can give different results on different rows. Thus, we can do with our data two

---

[3] To separate the two factors we should have asked the operators to take the data with a scheme that would have made the two parts independent. It is a commonly adopted process in the field of mass measurements, where, when comparing a mass standard, indicated by the symbol A, and a mass to be calibrated, indicated with the symbol B, the ABBA scheme is used for the weighing operations. We weight first the mass A, then B, then again B and finally A, so that the presence of a drift can be eliminated by linear operations of average. Other methods are given by the techniques of Design of Experiments (DoE).

types of analysis: one type, with a single factor under control, and another type that allows to manage the simultaneous presence of two factors In the first type of analysis, an example is done on the operator effects; but in the same way it could be possible to do it on the time effect alone. With our data no other type of analysis could be made. Note that the factor time can be analysed just because of the organization scheme used for the measurements: would the measurements be taken freely at any time, that factor could not be controlled. This shows that, in order to apply the ANOVA, we have to decide before doing the experiment what factors may be worth checking and organise the measurements in a suitable manner. Schemes of organization of the experiment and an extended discussion of this method can be found in specific texts[4]. Here only two cases of common use are examined, corresponding to the data of Table 8.1: analysis of a single or of two factors under control.

8.2.1 Analysis of variance for a single factor under control

A very common situation requires the evaluation of a single factor. Let's take the example of the operator's behaviour using the data of Table 8.1 corresponding to the subdivision scheme organized in data groups, of which the average is calculated, as reported in Table 8.3, where:

x_{ij} = i[th] measured data of the jth group

$\bar{x}_{\bullet 1}$ = averages of the jth group

$\bar{x}_{\bullet\bullet}$ = grand average of all data

In the exemplified case, the data changes can be subdivided into variations due to the operators (factor that we want to analyse) and natural variations due to random effects (factors different from the operators).

It is clear that systematic variations due to the operators are strongly present in the average values obtained by all the operators: the variance of these averages will strongly contain the effect of operator differences. Considering the experimental data divided into different groups for the various conditions of the parameter under control (in the example, in groups of data each characterized by a different operator).

We understand the traditional name "variance between groups" and the symbol s_B^2 assigned to it, as it takes into account the differences between groups:

$$s_B^2 = ns_m^2 = n\frac{\sum_{J=1}^{k}\left(\bar{x}_{\bullet j} - \bar{x}_{\bullet\bullet}\right)^2}{k-1} = \frac{SS_B}{k-1}$$

it is made using the sum of the squares of differences between the average values of each group and the global average value (Sum of Squares between groups, SS_B).

[4] D. C. Montgomery, G. C. Runger. "Applied Statistics and Probability for Engineers", Chapter 13

We shall, then, determine an estimation of the variance that does not contain the effect of the differences between the groups (in our case between operators). It is evident that, if we calculate the variance of the data of a single operator, we will get a value independent from the difference against other operators. It is called “variance within groups”, indicated with the symbol s_W^2.

Tab. 8.3 Data scheme for a single factor under control

Conditions	Group 1	...	Group j	...	Group k	Grand average
1	x_{11}	...	x_{1j}	...	x_{1k}	
...	...	...	...	...	...	
i	x_{i1}	...	x_{ij}	...	x_{ik}	
...	...	...	...	...	...	
n	x_{n1}	...	x_{nj}	...	x_{nk}	
Group average	$\bar{x}_{\bullet 1}$	...	$\bar{x}_{\bullet j}$	...	$\bar{x}_{\bullet k}$	$\bar{x}_{\bullet\bullet}$

To make this information more robust, finally, the operation is repeated for the other operators and it becomes the average of the estimates thus obtained for each group:

$$s_W^2 = \frac{1}{k}\sum_{j=1}^{k}\frac{\sum_{i=1}^{n}\left(x_{ij}-\bar{x}_{\bullet j}\right)^2}{n-1} = \frac{SS_W}{k(n-1)}$$

where the part:

$$\frac{\sum_{i=1}^{n}\left(x_{ij}-\bar{x}_{\bullet j}\right)^2}{n-1}$$

is the variance of the data within the jth group and the sums of the squares of the differences respect the average values of each group represent the Sum of Squares Within groups (SS_w).

To evaluate the presence of a systematic factor variable between groups we can do the ratio of the variance estimation containing the factor under control to the one that does not contain it, and compare this ratio with the upper limit[5] F_U of the confidence interval of the corresponding Fisher's distribution.

In case of a single factor under control, it has been possible to reach the formulation of an estimation of the variance that contains the examined factor and another estimation that does not contain it.

[5] The ratio of the variances made in this way is polarized. The actual presence of the factor in question makes the expected ratio greater than 1; so it is only interesting to compare the upper limit of the confidence interval.

For the case of two factors under control, it is more complex to find step by step a formulation of the variance unaffected by the effects of possible systematic factors, but it is possible to achieve this with a mathematical development[6].

We continue to examine the case of a single factor under control. We can subdivide the differences between the individual data and the overall average in differences between individual data and the average values of the groups and in differences between the average values of the groups and the overall average:

$$x_{ij} - \bar{x}_{\bullet\bullet} = \left(\bar{x}_{\bullet j} - \bar{x}_{\bullet\bullet}\right) + \left(x_{ij} - \bar{x}_{\bullet j}\right)$$

we can do now the sum of squares of the differences; recalling that the sum of differences from mean value is identically null, we can cancel the sum of products of first degree terms:

$$\sum_{J=1}^{k}\sum_{i=1}^{n}\left(x_{ij} - \bar{x}_{\bullet\bullet}\right)^2 = \sum_{J=1}^{k}\sum_{i=1}^{n}\left[\left(x_{ij} - \bar{x}_{\bullet j}\right) + \left(\bar{x}_{\bullet j} - \bar{x}_{\bullet\bullet}\right)\right]^2 =$$
$$= \sum_{J=1}^{k}\sum_{i=1}^{n}\left(x_{ij} - \bar{x}_{\bullet j}\right)^2 + 2\sum_{J=1}^{k}\sum_{i=1}^{n}\left(x_{ij} - \bar{x}_{\bullet j}\right)\left(\bar{x}_{\bullet j} - \bar{x}_{\bullet\bullet}\right) + \sum_{J=1}^{k}\sum_{i=1}^{n}\left(\bar{x}_{\bullet j} - \bar{x}_{\bullet\bullet}\right)^2 =$$
$$= \sum_{J=1}^{k}\sum_{i=1}^{n}\left(x_{ij} - \bar{x}_{\bullet j}\right)^2 + n\sum_{J=1}^{k}\left(\bar{x}_{\bullet j} - \bar{x}_{\bullet\bullet}\right)^2$$

the contributions traditionally have the shape:

$$SS_{TC} = \sum_{J=1}^{k}\sum_{i=1}^{n}\left(x_{ij} - \bar{x}_{\bullet\bullet}\right)^2 = \sum_{J=1}^{k}\sum_{i=1}^{n}\left(x_{ij} - \bar{x}_{\bullet j}\right)^2 + n\sum_{J=1}^{k}\left(\bar{x}_{\bullet j} - \bar{x}_{\bullet\bullet}\right)^2 = SS_W + SS_B$$

and the sum of squares within groups:

$$SS_W = \sum_{J=1}^{k}\sum_{i=1}^{n}\left(x_{ij} - \bar{x}_{\bullet j}\right)^2$$

and the sum of squares between groups:

$$SS_B = n\sum_{J=1}^{k}\left(x_{\bullet j} - \bar{x}_{\bullet\bullet}\right)^2$$

which are the numerators of the two variances s_W^2 and s_B^2 considered before.

The test can be organized directly calculating these variances or following the traditional scheme, which exploits the fact that the total sum of squares and total degrees of freedom (D.o.F.) can be subdivided among different factors, but do not change.

---

[6] The analysis of variance is a very common, but very specialized treatment, so it is appropriate to use diagrams and formulas given in specialized texts.

Therefore, a difficult contribution can be calculated as difference considering the total constant and calculating only the other easy ones. The most difficult contribution to calculate, the one that does not contain the factors under control, is obtained by difference.

In the shown example the total D.o.F. are 49 (number of data minus 1). The D.o.F. of the variance of the 5 average values are, for the same reason, 4. Therefore, by difference, the D.o.F. of the variance within the group s_W^2 will be 49-4 = 45.

This corresponds to what is obtained in a more complex way recalling that s_W^2 was calculated from the variances of 10 data per group, then with 9 D.o.F., taken 5 times, then 9·5=45.

On the basis of this principle the ANOVA table is calculated as shown in the scheme given in Table 8.4 and having the results shown in Table 8.5.

Tab. 8.4 Scheme of the ANOVA table for a single factor under control

Variation cause	D.o.F	Sum of squares	Variance	Variance Ratio	Fisher F
Due to the factor under control	k - 1	$SS_B = n\sum_{J=1}^{k}\left(\bar{x}_{\bullet j} - \bar{x}_{\bullet\bullet}\right)^2$	$s_B^2 = \frac{n\sum_{J=1}^{k}\left(\bar{x}_{\bullet j} - \bar{x}_{\bullet\bullet}\right)^2}{k-1}$	$F_{\exp} = \frac{s_B^2}{s_W^2}$	F_U
Due to random errors	$k(n$-1)	$SS_W = \sum_{J=1}^{k}\sum_{i=1}^{n}\left(x_{ij} - \bar{x}_{\bullet j}\right)^2$	$s_W^2 = \frac{\sum_{J=1}^{k}\sum_{i=1}^{n}\left(x_{ij} - \bar{x}_{\bullet j}\right)^2}{k(n-1)}$		
Total	kn - 1	$SS_{TC} = \sum_{J=1}^{k}\sum_{i=1}^{n}\left(x_{ij} - \bar{x}_{\bullet\bullet}\right)^2$			

Tab. 8.5 Results of the calculation of the ANOVA for the experimental data of table 8.1

Variation cause	D.o.F	Sum of squares	Variance	Variance Ratio	Fisher F
Due to operators	*4*	1,32E-04	3,31E-05	6,4	1.6
Due to random errors	*45*	2,33E-04	5,17E-06		
Total	*49*	3,65E-04			

Note the formulas of the D.o.F., which corresponds to what said above: the total D.o.F. are equal to the total number of data minus 1 (kn–1).

The variance between the groups is determined with k averages, with k-1 D.o.F. and eventually the variance within the groups, calculated for k groups of n data, will have for each group n–1 D.o.F. and, therefore, in total, $k(n-1)$ D.o.F.

The latter value can also be calculated subtracting the D.o.F. of the variance between the groups from the total D.o.F.:

$$kn-1-(k-1)=kn-1-k+1=k(n-1)$$

The value of the variance ratio must be compared with the upper limit of Fisher's confidence interval F_U, calculated at the level of confidence required using the D.o.F. of the numerator and denominator.

It is found from Fisher's distribution that, at the confidence level of the 80% with 4 D.o.F. (numerator) and 45 (denominator), the upper limit is $F_U=1.6$.

Since the value of the variance ratio is F_{exp} = 6.4, higher than the upper limit, therefore outside the confidence interval, it can be affirmed, with a risk of error of 20%, that the differences between the operator's averages are systematic.

Note that even the analysis of variance is helpless in front of the superposition of the two effects of drift and differences between operators, indicated as cause of the systematic differences.

ANOVA cannot face two superimposed effects working in such a way that the variability on the operators is the total of the drift (the first operator represents a condition of minimum drift and the last one of maximum drift) and the question posed is on the variability connected to the operators get, for the numbers being, a correct answer to a wrong question. A better question to which is possible to get an answer applying ANOVA test, should involve operators and time together.

8.2.2 Analysis of the variance for two factors under control

The experimental data given above allow to control both the factor corresponding to the different columns (groups) and to the different rows (conditions).

Let's resume the collected data according to the scheme of table 8.6, where:

x_{ij} = measurement data of the jth data, acquired in the ith situation

$\bar{x}_{\bullet j}$ = average of the jth group data

$\bar{x}_{i\bullet}$ = average of the data acquired in the ith condition

$\bar{x}_{\bullet\bullet}$ = average of all the data

In this case the variability due only to the random effects must be obtained subtracting the variability effects of the two factors under control.

We evidence these factors, described by the differences between the averages for the columns and the overall average and between the averages for the rows and the overall average:

$$x_{ij}-\bar{x}_{\bullet\bullet}=\left(\bar{x}_{\bullet j}-\bar{x}_{\bullet\bullet}\right)+\left(\bar{x}_{i\bullet}-\bar{x}_{\bullet\bullet}\right)+\left(x_{ij}-\bar{x}_{\bullet j}-\bar{x}_{i\bullet}+\bar{x}_{\bullet\bullet}\right)$$

This leads, making the sum of the squares similarly to what seen before and eliminating the parts that lead to sums of products of deviations from the average, identically null, to:

$$\sum_{j=1}^{k}\sum_{i=1}^{n}\left(x_{ij}-\bar{x}_{\bullet\bullet}\right)^2=\sum_{j=1}^{k}\sum_{i=1}^{n}\left[\left(\bar{x}_{\bullet j}-\bar{x}_{\bullet\bullet}\right)+\left(\bar{x}_{i\bullet}-\bar{x}_{\bullet\bullet}\right)+\left(x_{ij}-\bar{x}_{\bullet j}-\bar{x}_{i\bullet}+\bar{x}_{\bullet\bullet}\right)\right]^2$$

$$\sum_{j=1}^{k}\sum_{i=1}^{n}\left(x_{ij}-\bar{x}_{\bullet\bullet}\right)^2=n\sum_{j=1}^{k}\left(\bar{x}_{\bullet j}-\bar{x}_{\bullet\bullet}\right)^2+k\sum_{i=1}^{n}\left(\bar{x}_{i\bullet}-\bar{x}_{\bullet\bullet}\right)^2+\sum_{j=1}^{k}\sum_{i=1}^{n}\left(x_{ij}-\bar{x}_{\bullet j}-\bar{x}_{i\bullet}+\bar{x}_{\bullet\bullet}\right)^2$$

Examining all the different contribution, we can see that $n\sum_{j=1}^{k}\left(\bar{x}_{\bullet j}-\bar{x}_{\bullet\bullet}\right)^2$ represents the numerator of the variance of the averages per column, while $k\sum_{i=1}^{n}\left(\bar{x}_{i\bullet}-\bar{x}_{\bullet\bullet}\right)^2$ represents the numerator of the variance of the averages per row. Eventually, $\sum_{j=1}^{k}\sum_{i=1}^{n}\left(x_{ij}-\bar{x}_{\bullet j}-\bar{x}_{i\bullet}+\bar{x}_{\bullet\bullet}\right)^2$ is the numerator of the variance unaffected by the suspected systematic factors, therefore considered an estimation of the variance of the random effects involved.

Tab. 8.6 Data Scheme for two factors under control

Factor 2	**Factor 1**					**Factor 2 average**
	Group 1	...	Group j	...	Group k	
Cond. 1	x_{11}	...	x_{1j}	...	x_{1k}	$\bar{x}_{1\bullet}$
...	...	...	...	...	...	...
Cond. i	x_{i1}	...	x_{ij}	...	8	$\bar{x}_{i\bullet}$
...	...	...	...	...	...	...
Cond. n	x_{n1}	...	x_{nj}	...	x_{nk}	$\bar{x}_{n\bullet}$
Factor 1 average	$\bar{x}_{\bullet 1}$	...	$\bar{x}_{\bullet j}$	...	$\bar{x}_{\bullet k}$	$\bar{x}_{\bullet\bullet}$

The calculation can be done according to the scheme of table 8.7, where for the D.o.F. we easily have the terms of the total variance and the first two contributions, respectively calculated with k data (therefore k–1 D.o.F.) and with n data (then n–1 D.o.F.). The D.o.F. of this contribution can be easily obtained by difference compared to the total:

$$kn\ 1\ (k-1)\ (n-1) = kn\ 1 - k + 1 - n + 1 = (k\ 1)(n\ 1)$$

The corresponding results are shown in Table 8.8. In this case, we must evaluate with Fisher's distribution two different conditions, always at 80% confidence level: the first with 4 D.o.F. at the numerator and 36 at the denominator, the second with 9 D.o.F. at the numerator and 36 at the denominator. The corresponding values of F_U are 1.58 and 1.46, to be respectively compared with the values of F_{exp} 6.92 and 1.41. We can say, with a risk of error of the 20%, that the differences between the averages of the operators are caused by systematic factors, while nothing can be said for the differences due to the sequence of measures.

Tab. 8.7 ANOVA table with two factors under control

Variation cause	D.o.F	Sum of squares	Variance	Variance Ratio	Fisher F
Due to factor 1 (Columns)	*k* - 1	$SS_{Bc} = n\sum_{J=1}^{k}\left(\bar{x}_{\bullet j} - \bar{x}_{\bullet\bullet}\right)^2$	$s_{Bc}^2 = \frac{SS_{Bc}}{k-1}$	$\frac{s_{Bc}^2}{s_W^2}$	F_{U1}
Due to factor 2 (Rows)	*n* - 1	$SS_{Br} = k\sum_{i=1}^{n}\left(\bar{x}_{i\bullet} - \bar{x}_{\bullet\bullet}\right)^2$	$s_{Br}^2 = \frac{SS_{Br}}{n-1}$	$\frac{s_{Br}^2}{s_W^2}$	F_{U2}
Due to random errors	(*k*-1) (*n*-1)	$SS_W = \sum_{J=1}^{k}\sum_{i=1}^{n}\left(x_{ij} - \bar{x}_{\bullet j} - \bar{x}_{i\bullet} + \bar{x}_{\bullet\bullet}\right)^2$	$s_W^2 = \frac{SS_W}{(k-1)(n-1)}$		
Total	*kn* - 1	$SS_{TC} = \sum_{J=1}^{k}\sum_{i=1}^{n}\left(x_{ij} - \bar{x}_{\bullet\bullet}\right)^2$			

Tab. *8.8 Results of the ANOVA table with two under control factors*

Variation cause	D.o.F	Sum of squares	Variance	Variance Ratio	Fisher F
Due to the factor 1 (Columns)	*4*	1.32E-04	3.31E-05	6.92	1.58
Due to the factor 2 (Rows)	*9*	6.06E-05	6.73E-06	1.41	1.46
Due to random errors	*36*	1.72E-04	4.78E-06		
Total	*49*	3.65E-04			

8.2.3 Comments on the analysis of variance

The analysis of variance has some advantages over the hypothesis tests initially described. These benefits are related to the ability to selectively isolate some contributions of possible influence factors and to determine their significance, allowing the application to more than a single factor at a time. Note that even the variance analysis cannot distinguish between two overlapped factors, in our case the sequence of the measures and the sequence of the operators. The analysis of the variance fails to highlight the effect of the drift shown by the regression. This is probably because, for ANOVA, the drift is segmented into parts, i.e. ANOVA visibility is on intervals much shorter than the total evolution of drift. On the contrary, it acts fully on variations assigned to the operators because, as underlined before, operator 1 feels the drift effect at minimum level while operator 5 feels it at the maximum level. In this way, the drift differences, within each operator data, lose of significance if compared to the higher differences that might be produced by the same drift on the total time, but assigned as effect to the operators. We see that, as with any statistical test, even for the analysis of variance it is necessary to look critically at the result. The regression has acted more strongly because it enjoyed the visibility of the complete data evolution.

It is convenient to operate in an iterative manner, since the identification and elimination of high disturbing effects, whether accident or systematic errors, makes statistical analysis subsequently applied more acute and selective.

Let's try, therefore, to correct for the drift evidenced by the regression, obtaining the data shown in Table 8.9. If we apply to the corrected data, both the hypothesis test on the averages and the ANOVA for a single factor under control (operators), we obtain the results shown in Tables 8.10 and 8.11.

Tab. 8.9 Results, expressed in millimetres and corrected from the drift, of the length measurements of a workpiece, done by five different operators. Control scheme for a single factor.

Time	Oper. 1	Oper. 2	Oper. 3	Oper. 4	Oper. 5	Grand Average
8	8,0061	8,0054	8,0064	8,0086	8,0037	
9	8,0073	8,0038	8,0044	8,0097	8,0059	
10	8,0058	8,0105	8,0030	8,0102	8,0095	
11	8,0038	8,0036	8,0050	8,0078	8,0093	
12	8,0043	8,0056	8,0025	8,0037	8,0046	
13	8,0081	8,0029	8,0029	8,0065	8,0044	
14	8,0098	8,0054	8,0071	8,0044	8,0093	
15	8,0075	8,0092	8,0062	8,0100	8,0042	
16	8,0047	8,0062	8,0067	8,0054	8,0042	
17	8,0099	8,0042	8,0059	8,0078	8,0033	
Average	8,0067	8,0057	8,0050	8,0074	8,0058	8,0061

The hypothesis test on the averages indicates, with the risk of error of 20%, significant differences for the operators 3 and 4. Even the analysis of variance shows, with the same risk of error, that the factor operators exceeds the limit F = 1.6. These results, really different from the previous ones, show that the presence of correlations has distorted the test answers. Thus we have attributed the drift's faults to the innocent operators.

Tab. 8.10 Determination of the confidence interval limits and average values obtained by the different operators on data corrected for drift

Parameters of the distribution of averages		**Operator**	**Average**
Number of data	10	1	8,0067
Average m	8,0061	2	8,0057
Standard deviation s_m	0,0007	3	**8,0050**
Confidence level $\mathcal{P}$	80%	4	**8,0074**
Lower limit	8,0052	5	8,0058
Upper limit	8,0071		

Tab. *8.11 Results of the ANOVA calculation for the experimental data of table 8.9*

Variation cause	D.o.F	Sum of squares	Variance	Variance Ratio	Fisher F
Due to operators	*4*	3,60E-05	8,99E-06	1,76	1.6
Due to random errors	*45*	2,29E-04	5,10E-06		
Total	*49*	2,65E-04			

An important tip is to randomize the factors that we want to analyse. A help is trying to detect the presence of "regularity" that point out the presence of non-random factors.

In our case the "clues" that can lead us to identify the real culprit are the shape of the GPN, that indicates the possibility of a drift in the measurements, and the almost regular succession of increasing averages values of the successive operators. Only the graphical presentation of all the experimental data and the analysis with the regression has allowed us to highlight the effect of the drift.

9 UNCERTAINTY OF TYPE B

9.1 *INTRODUCTION*

In Chapter 6 some statistical analysis methods of experimental data have been described. The theoretical basis of this evaluation, similar to what we see in many fields of statistical application, is to examine a statistical sample, a group, usually small, of the population (in technical field, "population" are measurements or products obtained in nominally identical way) deducing from it the information (average and variance) to be taken as estimation of the characteristics of the entire population. Two are the main limitations of this methodology:

- the large number of experimental data, necessary to have adequate estimates of the population's parameters;
- the requirement that the sample is representative of the characteristics of the population, that is, it contains information of all the factors that produce effects on the population.

The statistical methodology is, therefore, on a side expensive[1]due to the number of tests that it requires and, on the other side, it remains limited to experts, whom must carefully consider that all the important factors are contained in the examined group data. To overcome these limitations and to promote an efficient and economic management, the Guide on the expression of measurement's uncertainty (GUM) provides a schematization based on two different types of evaluation of factors contributing to the total uncertainty:

- an evaluation based on statistical methods, as we saw in Chapter 6, which leads to quantify the so called *Type A uncertainty contributions*;
- an evaluation based on not statistical methods, primarily due to the prior knowledge consolidated in technical-scientific literature, in regulations and in specific experience in the field in which the measurement takes place, leading to quantify the so called *Type B uncertainty contributions*.

In both cases, the aim is to find the value of the variance to be associated with each factor. Commonly is preferred to evaluate the standard deviation[2], because its value is comparable to the factor itself. Note that the two categories of contributions are not related to the subject of measurement or to the physical quantities involved, but only to the method used for the evaluation. Moreover they are not related to the systematic or random characteristic of the errors that correspond to the contributions of uncertainty.

---

[1] Cheaper than an unorganized survey of the population's characteristics, but very expensive compared to the use of information already consolidated by the experience.

[2] The GUM, introducing non-statistical method for evaluating uncertainty, avoids when possible the use of statistical terms, therefore names the standard deviation with the term *standard uncertainty*, for which uses the symbol u at the place of s used in statistics for standard deviation.

9.2 MEASUREMENT AND UNCERTAINTY

Let's examine the case of a length measurement of a product, considering a mathematical model of the measurement that takes into account the effects of temperature. The length varies according to the temperature.

As required by ISO[3], the length value to be used in industrial field (for example in technical drawings) is the one that the workpiece assumes at a temperature of 20°C. The length to be determined is, therefore, L_{20}, from the length L_m measured at temperature T.

$$L_{20} = L_m\left[1 + \alpha(20 - T)\right] \tag{9.1}$$

The measurement result is affected by many factors, some of which are negligible compared to the level of uncertainty that we want to obtain. It is possible to decide whether a factor is negligible only after having evaluated its effect. Moreover, if a factor is negligible, it is difficult to suspect his influence.

This leads to a vicious cycle that can be solved only according to the experience and the research. The first indicates the factors for which a significant effect is known on the basis of previous results. The second examines carefully the measurement results and evaluates whether they and their dispersion are justified by the factors taken into account, or whether it is necessary to identify the reasons for any discrepancies in the presence of other factors not yet considered.

As evident, the guide of the experience is indispensable for effectively managing a large part of factors, often directly eliminated as negligible.

This is particularly useful for increasing the efficiency of the measurement, which would have unsustainable costs if each influence factor had to be evaluated by statistical analysis. Analysis that is not, in general, heavy for the amount of calculations involved (a well-conceived spreadsheet allow to do routinely applications, even complex, in the short time necessary for data input), but certainly very expensive for the high number of measurements required for providing a sufficient[4] number of measurement replications for the statistical analysis.

In practice, in working conditions, we can perform only a limited number of tests. Some factors can be adequately evaluated with an acceptable number of replications, therefore can be statistically evaluated. Other factors produce, on the data collected, so small effects that are not detectable and would require a wider experimental basis, frequently unattainable for economic reasons. For others, the measurement is so complex that it cannot be made even once. We must not forget those factors that in the possible measurement replications are not affected by any variation, so it is not possible to highlight their effect.

---

[3] ISO 1:2002 -- Geometrical Product Specifications (GPS) -- Standard reference temperature for geometrical product specification and verification

[4] As seen in section 5.5.9 about the distribution of Pearson, for determining an estimate of the variance with an acceptable uncertainty it is necessary to have at least 30 data, and this means a high cost.

The example of length measurement contains many of the described conditions: the measure of L_m can be replicated several times and these replications, with their variability, put in evidence those factors involved in the measurement (contact force of the probe, state of vibration during the measurement, irregular shape of the measurand). Even the temperature will have its effect, but, if the replications are done in a short time interval, the dispersion of the obtained values will not contain any information on the effect of temperature variation, possibly wider in longer periods (along the day or during the seasonal variations[5]).

The coefficient of thermal expansion, however, is not subject to any variation in a normal test, since in replications the material of the measurand does not change. In general, also the instrument used to measure L_m will be just one, so the effect of its characteristics and its systematic errors will not contribute to the variability of the measurement results.

In practice, the variability field of the replicated results represents only a part of the measurement uncertainty, since, being done in a limited time, it is only partially affected by the possible temperature variation. Moreover, it is neither affected by the difference of the thermal expansion coefficient of the workpiece material, against the significant difference from the data taken in handbooks, nor by possible constant systematic errors of the measuring instrument.

This example shows that for some factors involved in a measurement it is not possible, practically, to plan an experiment which involves the natural variability of all the influence factors, such as to allow a statistical evaluation of their effects. There are consolidated information on some of the factors described before:

- the variability of instrument performances in typical working conditions, including the effect of constant systematic errors, are well evaluated by the tolerances reported in Technical Standards[6];
- the amplitude of typical diurnal variations in a working environment are known from experience;
- the effect of the reading resolution is clearly defined by the configuration of the scale, even for the metric instruments for which it is difficult to find out its effect with a reproducibility test.

---

[5] It is very difficult, for practical reasons, to collect a group of data that completely represent the variability of a measurement. In fact, in the case of testing activities that go on for months or years, the qualification phase of the process, taking into account seasonal variations in temperature, should also be continued for at least one year.

[6] For instance ISO 3611:2010 (Geometrical product specifications (GPS) -- Dimensional measuring equipment: Micrometres for external measurements -- Design and metrological characteristics) establish a tolerance of ±4μm for micrometres having a range of 25 mm. This information is in many cases better (it contains also the effect of the typical systematic constant error) than the information given by a reproducibility test.

We can see, therefore, the two limits of the management of type A uncertainty: the elevated number of tests and the difficulty to really have all the possible sources of involved variability.

Using the support of previous experience in all its evidences has the following advantages:

- it is possible to consider factors not contained in the replicated data;
- some of the influencing factors can be neglected on the basis of experience;
- factors evolving in long periods of time, for instance the temperature for external tests, can be considered on the basis of climatic tables;
- the typical metrological characteristics of instruments can be known on the basis of data sheets.

The considerable information content available on the basis of experience, which frequently is more reliable than the uncertainty evaluation made on the basis of a few experimental data, is evident.

In conclusion, we should emphasize four concepts:

- uncertainty factors to be taken into consideration are contained in up to date measuring procedures; for routine work it is not necessary to analyse other factors;
- some of the influence factors in possible testing conditions produce limited effects as compared with the working conditions; therefore cannot be evaluated using statistical methods, so they must be managed with other methods (type B uncertainty contributions);
- when past experience allows to quantify the contributions of several factors (type B uncertainty contributions), this way of analysis should be preferred, since, in general, it is more robust in its conclusions and extremely less expensive.
- the statistical evaluation (type A uncertainty contributions) should be used only when the data is already available in sufficient quantities or when alternative ways are not viable.

From this discussion it is should be evident that, even if calculations for evaluating uncertainty are not complicated, decisions on the factors to be evaluated and the variability to assign to each factor requires a deep experience of the involved topics. We underline again the importance of a good connection to process technicians, but we can provide some very helpful tools for solving problems that technicians do not have. One is the statistical analysis seen before, for evidencing if there are problems connected with systematic factors (normality tests), which factors are likely to be the causes of the problems (hypothesis tests and ANOVA) and if they can be corrected with a mathematical model (regression). An additional tool is the uncertainty analysis: we shall be able to use it efficiently, as described in ISO 14253-2, with an important and useful method, so called PUMa method (Procedure for Uncertainty Management). The problems underlined before are faced in a very practical way: as said before, the problem is not the calculation itself, but how to select the influence factors and the amount of their associated variability; the real problem is to assign variability, because the number of factors does not complicate the procedure. The PUMa method consists to insert freely any factor evidenced by the technical part, when necessary split it in the relevant metrological characteristics, to assign known variability and, when the variability of one factor is not known, just put, with the help of the technical part, a reasonable value. Thereafter calculations are immediate and they will show which factors are really important and which are negligible; for the most important factors, whose variability was given by guess, it will be necessary to get better information while for the negligible ones there will not be necessary to do other work.

9.3 EXAM OF SOME UNCERTAINTY CAUSES

The distinction between Type A and Type B uncertainty is related to the evaluation procedures and not on the conditions of the examined factors. The GUM defines as Type A uncertainty contributions "those evaluated using statistical methods"; as Type B "those assessed by other methods." The same factor could be evaluated one time replicating measurements and calculating the variance, and another time just taking the relevant information from the previous experience (e.g. from a Standard) . When the variance is used, that is a statistical method, the contribution is of A type; when the information is taken from the previous experience, therefore assessed by a non-statistical method, the contribution is of B type.

Consider the common situation of an instrument calibrated, with repeated measurements, in a metrological laboratory with controlled temperature within ± 0.5 °C and used in a workshop with a temperature variability of ± 5 °C. This requires a management with both methods:

- calibration data have replication and statistical analysis, giving therefore Type A contributions; but the effect of a large variation of temperature is not involved. We can see that, for temperature, we have an information of repeatability;
- for assessing the variability of our results taken in the working conditions we need an information of reproducibility. We can take the information on the temperature sensitivity of the instrument in the data sheet, therefore information obtained from the previous experience, thus leading a Type B contribution.

Note, however, that if in the working conditions it is possible to qualify our instrument, that is asses its reproducibility in the working condition with a sufficient number of test on a stable measurand (tests that shall explore completely the possible temperature field), we can get again a type A contribution, but at the cost of many tests (at least 15, better 30), presumably on a long time to allow temperature to go throughout its natural variability interval.

The logical categories to be adopted are systematic effects and random effects. With replicated measurements we can assess the random variability. The point is that we are interested in the variability of the effects, which are proportional to the variability of the relevant factors. If we test an instrument in a temperature controlled chamber and we do not notice any effect of temperature, we cannot say that the instrument is not sensitive to the temperature. What we need is to have the response of our instrument to the common variation in working conditions. This recalls us the definitions of repeatability and reproducibility. So, if we have an incomplete testing condition, like when common variations of working conditions are not explored, a Type A contribution is inadequate. For the same reason it is not possible to get systematic constant effects with replications, as they do not produce variability in replicated results.

In all that cases Type B contributions are very useful.

Note that it seems really different the goodness of Type A and Type B contributions: Type A is based on a rigorous theoretical basis and on specific experimental data, while Type B consists just to take information somewhere without any control. For instance we can consider the following case: we have to measure with a micrometer the thickness of a metal sheet which is near the ceiling of our workshop. Type a procedure requires that we ask our operator to go up there and to take at least 30 measurements to be statistically analysed. For Type B we just present the work to an expert operator and ask him: “In your opinion, viewing the difficulty of measurement, how much could be the variability of the result?". In few minutes we get our answer.

The common sense would consider this Type B information very poor compared with the Type A described before, but the guess of an expert operator could even be better than the statistical information obtained with 30 data. GUM gives on these clear warnings.

> ***E.4.3*** *Evaluations based on repeated observations are not necessarily superior to those obtained by other means.....*
>
> *....Type A evaluations of standard uncertainty are not necessarily more reliable than Type B evaluations, and that in many practical measurement situations where the number of observations is limited, the components obtained from Type B evaluations may be better known than the components obtained from Type A evaluations.*

We can examine two very common cases:

- the variability to be assigned to numerical information (data taken in handbooks, data-sheet, etc.);
- the variability to be assigned to the bias declared for measuring instruments.

The first case is really very frequent: one gets a data being just a number, without accessory information of variability, like an uncertainty or a tolerance assessment. In this case the consideration to be done is that the number was written correctly: when we write a number corresponding to a physical quantity[7], we shall not write illusory figures, i.e. figures which are lower than the variability involved. We shall be careful because, unfortunately, the use

[7] ISO 80000-1:2009 -- Quantities and units -- Part 1: General

of more digits than necessary for transmitting the real information is quite common: one uses an instrument with the resolution of 0.01 mm, do some calculations and, as the calculator is set to give six decimal digits, get and transmit the result with 1 nm resolution! Nevertheless, people with a minimum of experience, good handbooks, data sheets and catalogues are keen in posing the correct number of significant digits: this means that the variability affects only the least significant digit of the number. But how many units of the least significant digit? We have seen for technical instruments that the variability is not just ± 1 unit of resolution, but frequently ± 2 units, sometimes ± 3 units; analogously assigning a variability of ± 2 units of the least significant digit can be considered as reasonable. A little more attention can be given to the numerical tables in handbooks, which can present numbers with the resolution of 2 or 5 units of the least significant digits (all numbers end with an even figure in the first case, with 0 or 5 in the second). In this case it can be considered reasonable to use as variability ± 1 unit of the relevant resolution.

The second case requires attention because the information on bias can arrive in two different ways:

a) we have a calibration certificate of the instrument, comprehensive the description of its input-output characteristic and the relative uncertainty (Type A) determined with the calibration operation. In this case, the corrections derived from the input-output characteristic shall be applied to the measurement results and the uncertainty stated in the certificate remains as an uncertainty contribution;
b) we only have a declaration of conformity to the relevant technical standard or the information on the accuracy class. In practice, we only know a variability interval (the tolerance prescribed by the technical standard or declared in the uncertainty class) in which the bias can be, or better, the combined bias and repeatability error can be; but, since the more severe effects are due to bias, penalizing a little the situation, we can consider it as totally due to bias. For example the ISO Standard ISO 3611:2010 on 25 mm range micrometres says that the maximum error shall be within ± 4 µm. So, if we are using a micrometre with declaration of conformity to ISO 3611, we know that its maximum error can be anywhere in the interval ± 4 µm. In this case, we take that interval as possible variability of our measurement results, but it do not correspond to the variability of measures, but to the variability of the characteristics of that type of instrument; therefore we must make an evaluation of the type B contribution.

9.4 Evaluation of the Type B uncertainty

This second situation is very common: the information on variability to be managed as type B contributions usually arrives as an interval of variability. For instance, as said before, we know that our instrument has an accuracy class of 0.1%, that is its maximum error is in the interval ± 0.1%; we guess that the temperature in our laboratory is between 18 °C and 28 °C, we get the number 1.2435 and we guess that the variability is ± 0.0002; we ask our operator about the possible variability of the measurement of the thickness near the ceiling of our laboratory, and he says that is really difficult to read the micrometre in that position, so this will increase the variability that could be about 5 μm.

We shall note that there are two different situations: in some cases the number gives the value of the half-interval, as can be recognised by the ± sign before the number, while for the temperature the information gives the total interval of 10 °C. But sometimes we cannot know the meaning of the given number: for instance the answer of the operator refers to the total interval or to the half-interval? The question is important, because both the situations are very common, even conventional: in the field of measurement, the convention is to give the half-interval without the sign ±; therefore, if our operator follows this convention, 5 μm are the semi interval. Note that this convention is general; therefore, in case of a maximum error of 4 μm or an expanded uncertainty of 2 μm, 4 and 2 represent the half-interval. Also the confidence interval is given in the same way and it is said conventionally to be 2σ (normal distribution at a confidence level of 95%). For all that cases variability intervals are twice the given number.

The convention is different in the Quality field. It is well known that the conventional confidence interval is 6σ (±3σ, normal distribution at a confidence level of 99.7%) and tolerances are given with the value of the total interval.

In conclusion, when the information received has the sign ±, no doubt that it is the half-interval. Tolerances, when given without the ± sign are the total interval. Expanded uncertainty, standard uncertainty, maximum error shall be given as half-interval without ± sign (e.g. this instrument has a maximum relative error of 0.1%); only when written together with the numerical value the sign ± shall be written (e.g. 25.7±1.3). If we are not sure about the used convention by who gives us the variability information, we have to ask if the number represents the half-interval or the total interval. GUM utilises the half-interval indicated generally with the symbol a.

The starting point for managing Type B contributes is, therefore, to be sure about the value a of the half-interval of variability. Thereafter, for proceeding in the classical statistical way of combining the effects summing the variances the approximate and conventional but very efficient way indicated by the GUM consists to pass from the half-interval a to the associate variance. Now, to pass from an interval of variability to a variance a hypothesis of the shape of the relevant statistical distribution is necessary. There is the approximation and the efficiency: GUM considers only three types of distribution: arcsine, uniform and triangular.

To decide which one shall be applied is simple: we shall take into account the character of factor under evaluation. It can have an oscillating characteristic (e.g. uncertainty produced by oscillations, vibrations, natural temperature variation for periods longer than one day, effects produced by rotations, including circularity tolerances). As seen in section 5.5.3, when the factor X is oscillating with a sinusoidal pattern in the interval $\pm a$, the associated distribution of probability is the arcsine distribution, also called U distribution, which has a variance equal to $a^2/2$. Once again, the calculation is very simple, but the important point that requires attention is to notice if the oscillating factor is felt as such or not by the measurement results.

Two examples, showing two common conditions of oscillating factors not affecting the results, can be useful.

The first example is the oscillating natural value of the temperature over 24 h: temperature has a minimum at about 4 o'clock in the morning and a maximum at about 4 o'clock in the afternoon, going with a nearly sinusoidal pattern in the period T of about 24 h. But if we perform our measurements only between eleven and twelve in the morning, our measurements do not feel the oscillation. We can say, in general, that when the time period in which is possible to have measurements is much shorter than the period of the oscillation, the oscillation effect does not affect the results.

The other example considers the condition of a time interval, used by the instrument for sensing and integrating the signal value, much longer compared with the oscillation period: the effect of the oscillation will not be felt by the result. This is the same situation of taking a lot of data for a time much longer than the oscillating period: if there is not a synchronization between oscillation and measurement that gives us results taken at the same phase of the oscillation, the average obtained do not feel the effect of oscillation.

In these two cases the influence factor is oscillating, but the results do not feel it, therefore the U distribution shall not be applied.

Oscillating factors are normally well known and well represented, as pertain to the variability effects, by a sinusoidal oscillation. But what about other types of factors? GUM is drastic: the alternative is a very expensive experiment for evaluating the distribution shape, but this is within Type A procedures, or ignorance, that involves Type B. Ignorance is just subdivided in two conditions: 1) nothing is known about the distribution; 2) there is a guess that the central values of the variability interval are more probable than the external ones. GUM assign to these two conditions a uniform distribution, for which the variance is $a^2/2$ (section 5.5.1) and a triangular distribution, for which the variance is $a^2/6$ (section 5.5.2), respectively.

The practical scheme is, therefore, very simple: we have to consider each factor producing a contribution of Type B and define its half-interval variability a; if the influence factor has an oscillating pattern, the equivalent variance is calculated as $u^2 = a^2/2$; if the influence factor is not oscillating, generally a uniform distribution is associated with a variance $u^2 = a^2/3$. Only in special cases, normally managed by experts of the process, a triangular distribution with a variance $u^2 = a^2/6$ is considered.

This simple procedure can be very useful for designing our process (measurement or production). As said in chapter 1, to take decisions we need to have information about variability, but at the design stage, we cannot have experimental data; therefore, we cannot perform a Type A analysis. It is at this stage that the use of the evaluation of Type B uncertainties presents the major advantages[8]. We must decide which factors are negligible, which ones can be considered simply as contributions of Type B and which to be subjected to a more detailed analysis with the possible determination of the contribution of Type A.

At this stage, it is appropriate to take into account the uncertainty contributions due to measuring instruments to be adopted, for which it is only known, from catalogues, the accuracy class or the compliance with the standard, an sufficient information for performing a Type B evaluation.

[8] A good preliminary calculation based on the evaluation of the type B uncertainty contributions can be completely adapted to the requirements of the regulations on Quality for the majority of industrial measurements, producing proper documentation with acceptable costs even in the small and medium enterprises.

10 THE COMBINED UNCERTAINTY

10.1 COMPOSITION OF THE UNCERTAINTY CONTRIBUTIONS

In the previous chapters we have seen how to evaluate the individual contributions of the various factors that produce the measurement uncertainty. What matters is not the single factor, but the combined effect of all of them together. This effect should be evaluated on a statistical base.

To do this we must consider that each measurement must be expressed as a function of different variables, some of them containing different factors. For instance, the length measurement at the temperature T described by equation 10.1 has the dependent variable L_{20} and the independent variables L_T, length measured at the temperature T, α, the thermal deformation coefficient, and T, the temperature during the measurement.

The variable L_T is measured, therefore its value is affected by the factors connected with bias, random errors and resolution of the measuring instrument.

$$L_{20} = L_T\left[1+\alpha\left(20-T\right)\right] \tag{10.1}$$

Even if it was not necessary to consider the effect of temperature, would always be present at least the resolution r of the used measuring instrument. Therefore, the value of length L, compared to the measured length L_T would be given by:

$$L = L_T \pm \frac{r}{2} \tag{10.2}$$

Even in the simplest case, we must consider that the magnitude to be measured Y (dependent variable) is a function of many quantities Xj (independent variables):

$$Y = G\left(X_1, \ldots, X_j, \ldots, X_q\right) \tag{10.3}$$

Such relation between the variables is translated into a similar one between the values of the quantities involved:

$$\bar{y} = G\left(\bar{x}_1, \ldots, \bar{x}_j, \ldots, \bar{x}_q\right) \tag{10.4}$$

In most cases the functions involved are quite regular to allow to be represented in an appropriate interval around the averages with a Taylor expansion limited to the first order terms[1].

---

[1] The development of a Taylor series arrested at the first order is sufficient to describe the effects of small variations of influence factors and of the accidental errors. In some cases, however, we can be in presence of strong interactions between factors, which may require the presence of higher order and mixed terms. (GUM, point 5.1.2).

If the independent variables x_j undergo a small variation, we can calculate the corresponding change of the dependent variable through the formula:

$$\delta y = \left(\frac{\partial G}{\partial x_1}\right)_{\bar{x}_1} \cdot \delta x_1 + \ldots + \left(\frac{\partial G}{\partial x_j}\right)_{\bar{x}_j} \cdot \delta x_j + \ldots + \left(\frac{\partial G}{\partial x_q}\right)_{\bar{x}_q} \cdot \delta x_q \tag{10.5}$$

Factors $c_j = \left(\frac{\partial G}{\partial x_j}\right)_{\bar{x}j}$ are the partial derivatives of the function G calculated by replacing the variables X_j with their average. They take into account how sensitive is the dependent variable y to the variations of the respective independent variables x_j, so they are called *sensitivity coefficients*.

If the relationship is expressed by a mathematical formula, we can make the algebraic or numerical derivation.

If it is expressed by a functional, for example by a conventional procedure such as the hardness measurement (to apply a force F for a time t, etc.), or by an empirical diagram, the value of the derivative is approximated by the value of the incremental ratio:

$$\left(\frac{\partial G}{\partial x_j}\right)_{\bar{x}_j} = c_j \approx \frac{G(\bar{x}_1, \ldots, \bar{x}_j + \Delta x_j, \ldots, \bar{x}_q) - G(\bar{x}_1, \ldots, \bar{x}_j, \ldots, \bar{x}_q)}{\Delta x_j} \tag{10.6}$$

where the increase should be chosen small enough not to be overly influenced by the local non-linearity, but large enough not to have too much uncertainty. In the case of strong non-linearity may be appropriate to use the following formula:

$$\left(\frac{\partial G}{\partial x_j}\right)_{\bar{x}_j} = c_j \approx \frac{G(\bar{x}_1, \ldots, \bar{x}_j + \Delta x_j, \ldots, \bar{x}_q) - G(\bar{x}_1, \ldots, \bar{x}_j - \Delta x_j, \ldots, \bar{x}_q)}{2\Delta x_j} \tag{10.7}$$

At this point we can describe what would happen in a series of n replicate measurements:

$$\begin{aligned}
\delta y_1 &= c_1 \cdot \delta x_{11} + \ldots + c_j \cdot \delta x_{1j} + \ldots + c_q \cdot \delta x_{1q} \\
&\ldots \qquad \ldots \qquad \ldots \\
\delta y_i &= c_1 \cdot \delta x_{i1} + \ldots + c_j \cdot \delta x_{ij} + \ldots + c_q \cdot \delta x_{iq} \\
&\ldots \qquad \ldots \qquad \ldots \\
\delta y_n &= c_1 \cdot \delta x_{n1} + \ldots + c_j \cdot \delta x_{nj} + \ldots + c_q \cdot \delta x_{nq}
\end{aligned} \tag{10.8}$$

where the subscript i indicates the differences from the average in the ith test. The system of equations thus obtained represents the linearized relationship between the dependent variable $\boldsymbol{Y}$ and the independent variables Xj. To continue the discussion is convenient to represent it in a matrix form:

$$\begin{Bmatrix} \delta y_1 \\ \cdot \\ \delta y_i \\ \cdot \\ \delta y_n \end{Bmatrix} = \begin{bmatrix} \delta x_{11} & \cdot & \delta x_{1j} & \cdot & \delta x_{1q} \\ \cdot & \cdot & \cdot & \cdot & \cdot \\ \delta x_{i1} & \cdot & \delta x_{ij} & \cdot & \delta x_{iq} \\ \cdot & \cdot & \cdot & \cdot & \cdot \\ \delta x_{n1} & \cdot & \delta x_{nj} & \cdot & \delta x_{nq} \end{bmatrix} \begin{Bmatrix} c_1 \\ \cdot \\ c_j \\ \cdot \\ c_q \end{Bmatrix} \qquad (10.9)$$

that is written in a more compact way:

$$\{\delta y_i\} = \lfloor \delta x_{ij} \rfloor \{c_j\} \qquad (10.10)$$

that easily allows the variance calculation of the dependent variable y:

$$s^2(y) = \frac{\sum_{i=1}^{n} (\delta y_i)^2}{n-1} = \frac{\{\delta y_i\}^T \{\delta y_i\}}{n-1} = \{c_j\}^T \frac{\lfloor \delta x_{ij} \rfloor^T \lfloor \delta x_{ij} \rfloor}{n-1} \{c_j\} \qquad (10.11)$$

We can also observe that the shape assumed by the product of the matrix of independent variable time its transposed:

$$\frac{\lfloor \delta x_{ij} \rfloor^T \lfloor \delta x_{ij} \rfloor}{n-1} = \frac{1}{n-1} \begin{bmatrix} \delta x_{11} & \cdot & \delta x_{i1} & \cdot & \delta x_{n1} \\ \cdot & \cdot & \cdot & \cdot & \cdot \\ \delta x_{1j} & \cdot & \delta x_{ij} & \cdot & \delta x_{nij} \\ \cdot & \cdot & \cdot & \cdot & \cdot \\ \delta x_{1q} & \cdot & \delta x_{iq} & \cdot & \delta x_{nq} \end{bmatrix} \cdot \begin{bmatrix} \delta x_{11} & \cdot & \delta x_{1j} & \cdot & \delta x_{1q} \\ \cdot & \cdot & \cdot & \cdot & \cdot \\ \delta x_{i1} & \cdot & \delta x_{ij} & \cdot & \delta x_{iq} \\ \cdot & \cdot & \cdot & \cdot & \cdot \\ \delta x_{n1} & \cdot & \delta x_{nj} & \cdot & \delta x_{nq} \end{bmatrix} =$$

$$= \frac{1}{n-1} \begin{bmatrix} \sum_{i=1}^{n} \delta x_{i1}^2 & \cdot & \sum_{i=1}^{n} \delta x_{i1} \delta x_{ij} & \cdot & \sum_{i=1}^{n} \delta x_{i1} \delta x_{iq} \\ \cdot & \cdot & \cdot & \cdot & \cdot \\ \sum_{i=1}^{n} \delta x_{i1} \delta x_{ij} & \cdot & \sum_{i=1}^{n} \delta x_{ij}^2 & \cdot & \sum_{i=1}^{n} \delta x_{ij} \delta x_{iq} \\ \cdot & \cdot & \cdot & \cdot & \cdot \\ \sum_{i=1}^{n} \delta x_{i1} \delta x_{iq} & \cdot & \sum_{i=1}^{n} \delta x_{ij} \delta x_{iq} & \cdot & \sum_{i=1}^{n} \delta x_{iq}^2 \end{bmatrix} = \lfloor s^2(x) \rfloor \qquad (10.12)$$

The matrix has on the main diagonal the variances $s^2(x)$ of the independent variables and out of it the summation of mixed products divided by the degrees of freedom (covariances). This is called variance-covariance matrix. In conclusion, knowing the behaviour of the independent variables, we can calculate the variance of the dependent variable:

$$s^2(y) = \{c_j\}^T \lfloor s^2(x) \rfloor \{c_j\} \qquad (10.13)$$

The variance-covariance matrix can be calculated exactly as the scheme proposed above: performing a series of measurements of the independent variables, calculating the deviations δx_{ij} from the averages (matrix $\lfloor \delta x_{ij} \rfloor$), from which it is easy to calculate $\lfloor \mathbf{s}^2(x) \rfloor$.

In the practice, we proceed to a further simplification.

In general, especially when considering the terms strongly random associated with the influence quantities, we can assume a complete independence of the corresponding variables.

So, if there is no specific indication of dependence between two or more independent variables, the covariances are considered null[2].

In this case, doing the calculations we have:

$$s^2(y)=\{c_j\}^T\left[s^2(x)\right]\{c_j\}=\frac{1}{n-1}\left[c_1 \; . \; c_j \; . \; c_q\right]\begin{bmatrix} \sum_{i=1}^{n}\delta x_{i1}^2 & . & 0 & . & 0 \\ . & . & . & . & . \\ 0 & . & \sum_{i=1}^{n}\delta x_{ij}^2 & . & 0 \\ . & . & . & . & . \\ 0 & . & 0 & . & \sum_{i=1}^{n}\delta x_{iq}^2 \end{bmatrix}\begin{Bmatrix} c_1 \\ . \\ c_j \\ . \\ c_q \end{Bmatrix}$$

$$s^2(y)=c_1^2\frac{\sum_{i=1}^{n}\delta x_{i1}^2}{n-1}+.+c_j^2\frac{\sum_{i=1}^{n}\delta x_{ij}^2}{n-1}+.+c_q^2\frac{\sum_{i=1}^{n}\delta x_{iq}^2}{n-1}$$

we obtain a general use formula:

$$s^2(y)=\sum_{j=1}^{q}\left(\frac{\partial G}{\partial x_j}\right)^2 s^2(x_j)=\sum_{j=1}^{q}c_j^2 s^2(x_j) \qquad (10.14)$$

that can be easily used. Two typical examples are given hereafter.

10.1.1 Uncertainty of the differences

Sometimes happen to determine the quantity $\boldsymbol{Y}$ required as difference[3] between two quantities X_1 and X_2. In this case the G function is given by:

$$Y=\mathrm{G}(X_2-X_1)$$

therefore, its variance is given by:

$$s^2(y)=\sum_{j=1}^{q}\left(\frac{\partial \mathrm{G}}{\partial x_j}\right)^2 s^2(x_j)=(-1)^2 s^2(x_1)+(1)^2 s^2(x_2)=s^2(x_1)+s^2(x_2)$$

---

[2] A typical case in which one must assume the presence of correlation is when we compose the data obtained with the same instrument. In this case, instead of using the 10.16 is convenient, although it is a bit pessimistic, to simply use the sum of the uncertainties.

[3] It's always better to avoid using the differences between measures. This is not to be confused with the method known as "Differential measurement method", which is very useful because it is organized in order to compensate by subtraction only the influence quantities, adding, instead, the contributions of the measured variable.

The value of the magnitude Y is smaller than at least one of the quantities X_1 or X_2, but its variance, then its uncertainty, is larger. The clearly unfavourable characteristic of this condition becomes even more evident considering the relative uncertainty, obtained as the ratio between the uncertainty and the measured value. The measurand is smaller of, one of the two parts really measured, but the relative uncertainties of these measurements, are at the state of the art, produce absolute standard uncertainties proportional to the really measured quantities. The measurand has a larger standard uncertainty (the combination of $s(x_1)$ and $s(x_2)$) and a smaller value, therefore a much larger relative uncertainty.

10.1.2 The uncertainty of the average

Another condition to which we should pay close attention regards the specific subject to the evaluation of uncertainty.

We decide to determine the tensile strength of a material. We select a series of samples and we submit them to the tensile test and, for the effects of random errors, we find a set of different results. The strength value to be assigned to the material will be the average of such data. Strictly speaking, if the value that we provide is the average, we must associate it to its uncertainty. So, given the variance s^2 of n experimental data, to have the variance s_m^2 of their average we shall divide it by n, as indicated by the central limit theorem (section 5.1.2). If we want to use that data to design a structure, the variance of the average is not a good indicator of the risk of rupture. In fact, the structure will not break for an average of the traction strength, but for the specific value of the piece of structure that can yield.

This example allows to highlight how we should always pay attention to the type of information that we need. Being easy to switch between the uncertainty of the average and the uncertainty of the individual data, this is applied whenever there is any discrepancy between the characteristics of the information that we receive (i.e. uncertainty on the average) and the one we really need (i.e. dispersion of the individual values).

We can see that the formula (10.14) leads to the relation already found with the central limit theorem. This happens in a more limited way because it does not say anything about the type of distribution of the average and only applies to the case where all the data are extracted from a single distribution characterized by a variance s^2. In this case we can apply the formula (10.14) taking into account that:

$$Y = \mathrm{G}\left(X_1, \ldots, X_j, \ldots, X_q\right) = m = \frac{1}{q}\sum_{j=1}^{q} x_j$$

therefore:

$$s^2(m) = \sum_{j=1}^{q}\left(\frac{\partial \mathrm{G}}{\partial x_j}\right)^2 s^2\left(x_j\right) = \sum_{j=1}^{q}\left(\frac{1}{q}\right)^2 s^2\left(x_j\right)$$

and being the data extracted all from the same distribution with variance $s^{2,}$ we have:

$$s^2(m) = \sum_{j=1}^{q}\left(\frac{1}{q}\right)^2 s^2(x) = q\left(\frac{1}{q}\right)^2 s^2(x) = \frac{s^2(x)}{q} \qquad (10.15)$$

This formula has a considerable importance, since the information on the conditions of variability can sometimes refer to averages. In that case, they are directly related to the variance $s^2(m)$. Very often, instead, they concern the individual values and, in this case, they are connected to the variance $s^2(x)$. It is clear that, if we use a variance instead of the other, we would do a serious mistake. In making the calculations it is often necessary to avoid mixing the two types of information. Therefore, as a rule, when we receive the information regarding the averages we use the formula (10.15) to transform the variance $s^2(m)$ in the corresponding variance $s^2(x)$, to be used for calculations. When, instead, the measuring result of which we want to evaluate the uncertainty is an average, we have to calculate the uncertainty of the average using the variance $s^2(m)$. For those factors which is known $s^2(x)$ using (10.15) we can pass to $s^2(m)$.

In practice, when receive the information of variability of an average, we must consider the *input multiplicity* n_d, which is the number of data averaged. When we use an average or we give an average as result, we must consider the *output multiplicity* n_r, which is the number of data averaged, being affected by random variability. Substituting in (10.15) to q one or the other multiplicity value, we fit the calculations to the conditions imposed by the averages used.

10.2 TABLE OF THE UNCERTAINTY BUDGET

In case of the evaluation of uncertainty, variances are obtained not only by Type A contributions, calculated with statistical procedure for which it would be justified to use the symbol s (estimated standard deviation σ) and s^2 for the variance, but also by Type B contributions, not based on statistical methods. The GUM prefer the concept of *standard uncertainty* (symbol u, variance u^2), which corresponds to standard deviation, for Type A contributions, and to equivalent standard deviation, for those of Type B. The equation of the uncertainty propagation obtained before (variance propagation) is, therefore, written in the form:

$$u^2(y) = \sum_{j=1}^{q}\left(\frac{\partial G}{\partial x_j}\right)^2 u^2(x_j) = \sum_{j=1}^{q} c_j^2 u^2(x_j) = \sum_{j=1}^{q} u_j^2(y) \qquad (10.16)$$

that uses the symbols c_j for the sensitivity coefficients and $u_j^2(y)$ for the contributions of each uncertainty factor to the variance of y.

To compare the contributions $u_j^2(y)$, the calculation of this formula can be organized in a table, which allows us to have a global view of the uncertainty budget, to understand the weight of the different contributions and to decide, if necessary, how to improve the measurement process. For this purpose it is very useful to use a scheme like the one represented in Table 10.1; the best way to show how to use it is a simple, but complete example, based, on the equation 10.1.

Columns x_j: symbols, values and notes about the independent variables Xj.

We can follow the mathematical model. The first variable is L_T, therefore we put its symbol in the first column, its value[4] in the second column and possible factors affecting it in the third column. Being a measured quantity, we can consider bias, resolution and precision, always remembering that it is not mandatory to have all the characteristics: if it is a comparison activity, bias is not important; if it is a stability monitoring, even precision is not important.

10.2.1 Process data

We have to look to the available information on variability involved in the process data. A typical base information is given hereafter. Practically, we use the process data available, adding, when necessary, some guessed reasonable values of influence factor, as said in PUMa method, to complete the necessary information. We perform the calculations and we look for critical points due to the guessed values, if there are, further analysis of critical points is required.

Data available are the following:

- bias of the measuring instrument is defined as the difference, against the reference value, of the average of 4 measurements, with an associated uncertainty of 3.7 μm declared by a Quality Service;
- repeatability is declared by the producer of the instrument as standard deviation of 1.3 μm;
- reading resolution of the instrument is 1 μm.
- measurements are taken during all 3 daily working periods, that is on the whole 24 hours, in a common working room without temperature control;
- the measurement procedure requires to perform 5 measurements and state their average as result;
- reproducibility was tested by 10 replications made by all operators, during all working periods, obtaining a variance of 10.7 μm²;
- temperature coefficient of the piece is $\alpha = 11 \times 10^{-6}$ °C^{-1} (a typical value for carbon steel, taken from handbook);
- temperature over a 24 hour time period is estimated to be 22 ± 5 °C.

10.2.2 Analysis of variables and factors

We have a typical set of information generally available. Starting with variable L_T, which is a measured quantity, we have to see the characteristics of the measuring instrument and, whenever possible, the reproducibility as effect of the working conditions.

---

[4] To avoid problems of measurement unit conversion on the result (difficult) it is better to avoid the use of multiples and submultiples in the input data, doing conversions at the beginning (simple).

We start with the bias, as it is important in this case. The relevant variability data is given as Uncertainty U[5]. We have seen that the uncertainty is obtained multiplying Student t time the standard deviation, therefore it is not possible to use directly the uncertainty information because the associated value can be very different. A measurement process is characterised by the standard deviation σ, but the associated uncertainty with the measurement convention is 2σ and with the Quality convention is 3σ, that is, in the same situation we can get very different values for the uncertainty, because different situations accept different risks of error. But to qualify the process, we need to know a unique value, variance or standard deviation. To transform the received Uncertainty information in standard deviation we need to know Student t, which is a function of the risk of error and the degrees of freedom (D.o.F.) adopted in the calculation of uncertainty (that is, the parameters used by who has calculated the expanded uncertainty). Different situations can happen:

- we receive the value of expanded uncertainty together with the accessory information of risk of error (or confidence level) and D.o.F. used for calculating it;
- who calculated the expanded uncertainty operated following the measurement conventions (confidence level 95% and a lot of D.o.F., well represented by the number 100), therefore he considered useless to give us the accessory information.

What we have to do is simple: if we receive the accessory information on confidence level and D.o.F. we use them, otherwise we use the conventional values, for one or both of them, when not declared.

In our situation we have a value of Uncertainty of the average of 4 measurements $U = 3.7$ µm declared by a Quality Service. Note that it means only that the uncertainty declared is the uncertainty of the average of 4 data (that is related to $\sigma_m = \sigma/4$) and not that the standard deviation used for evaluating the expanded uncertainty was calculated on 4 data. Therefore, in our case we have neither information on the D.o.F., nor on the confidence level. We shall put the given value of the expanded uncertainty in column 4, 99.7% as conventional value (Quality sector) of confidence level in column 5, and 100 as conventional value of the D.o.F. in column 6. Note that the heading of these columns have the proper symbols of expanded uncertainty U, confidence level P and D.o.F. ν. The table calculate automatically Student t (which is named coverage factor, with the symbol k by GUM) in column 7 and the relevant standard deviation in column 8.

[5] The GUM terminology is *Expanded Uncertainty*, symbol U, but GUM itself utilises the term *Uncertainty* for the same concept. Therefore it is better to use the complete terminology of "Expanded Uncertainty", but frequently information on it goes under the simple name of Uncertainty.

Table 10.1. Scheme for the Uncertainty budget

Mathematical model $L_{20} = L_T[1 + \alpha(20 - T)]$															
Variable x_j			Statistical					Non Statistical		Parameters estimated					
Symbol	***Value***	***Note***	U_j	P_{dj}	v_{dj}	k_{dj}	s_j	a_j	k_a	v_j	n_d	n_r	$u^2(x_j)$	$c_j = \Delta y / \Delta x_j$	$u_j^2(y)$
L_T	0.800074	Bias	3.7E-06	99.7%	100	3.0	1.2E-06		3	5	4	1	5.9E-12	1.0E+00	5.9E-12
		Res.		95.0%	100	2.0	0.0E+00	5.0E-07	3	100	1	5	1.7E-14	1.0E+00	1.7E-14
		Prec.		95.0%	100	2.0	3.3E-03		3	9	1	1	1.1E-11	1.0E+00	1.1E-11
α	1.1E-05	Table		95.0%	100	2.0	0.0E+00	2.0E-06	3	30	1	1	1.3E-12	-1.6E+00	3.4E-12
T	22	Estim.		95.0%	100	2.0	0.0E+00	5.0E+00	2	30	1	1	1.3E+01	-8.8E-06	9.7E-10
$y = L_{20}$	800.056	mm											Variance of y, $u^2(y)$		9.9E-10
													Standard uncertainty of y, $u(y)$		3.1E-05

The contribution of uncertainty for the factor bias is given, but we have to add the accessory data of D.o.F. and input and output multiplicity. Both columns 6 and 11 require the number of D.o.F.; what is the difference? We use column 6 only when in column 4 there is a value of expanded uncertainty and there we put the number of D.o.F. that we believe were used for calculating the given expanded uncertainty. In column 11 we have to put the value of D.o.F. we believe correct for the relevant factor. Again we can be in two different situations:

- we know the number of D.o.F. (e.g. we have received the value of expanded uncertainty together with the accessory information of the D.o.F. Or we have done 5 measurements used for evaluating the standard deviation, so we know that there are 4 D.o.F.)
- we do not know the number of D.o.F. (e.g. for any Type B contribution: how many D.o.F. shall we associate to the maximum error taken from an ISO standard? Or to information taken from an operator?). GUM says that Type B information are usually robust, therefore 100 D.o.F. shall be assigned. But certainly there is a difference in robustness from a data taken from an ISO standard or from a young operator; therefore, we prefer to have a confidence scale: 100 if we strongly believe in the variability information, 15 if we just believe in it and 30 (about the middle, as the scale is not linear) if we are between these two conditions. Moreover, bias characteristic has a peculiarity: it is evaluated by calibration and, looking on standard documents describing calibration procedures, we can see that usually the requirements are to repeat measurement very few times, sometimes even locally. In some cases it could be possible to evaluate the bias on the base of only three or five measurements; therefore, if for the bias is not declared the number of D.o.F., a precautionary assignment of five D.o.F. is suggested[6]. Therefore, in our case, we put 5 D.o.F.

Next step multiplicity. As said before, in principle it is not complicate. Recalling the equation 5.17, here reported as 10.17, it is clear that we must be very careful to distinguish the variance σ^2 of the individual data from the variance σ_m^2 of the average of q data.

$$\sigma_m^2 = \frac{\sigma^2}{q} \tag{10.17}$$

It is very easy to pass from one value to the other; the real difficulty consists in understanding when a variability information that we receive refers to the average (input multiplicity $\boldsymbol{n_d}$), and when, doing the average of some data or results, we involve random factors for which equation 10.17 is valid (output multiplicity $\boldsymbol{n_r}$).

[6] Always within the spirit of PUMa method: we try, and if the calculations show that it is a critical factor, we can ask some more information.

For the input multiplicity it is very important the description of the variability information that we receive. One typical mistake is confusing the number n of data used for evaluating the variance (or the standard deviation) with the number q of data averaged. The number n allows only to evaluate the D.o.F., while q is used for defining the input multiplicity n_d, and the output multiplicity n_r. In the case of the example is well described the fact that the variability information on the bias is “the expanded uncertainty of the average of 4 data”, that is, 4 is the number of data averaged and, therefore, the input multiplicity $n_d = 4$.

Looking at the procedure, we can see that it is required to carry out 5 measurements to be averaged for giving the result; therefore also output multiplicity is possible.

For output multiplicity the situation is more complicated because we have to go deep in the matter for understanding if, even in presence of an averaging, equation 10.17 can be applied or not. The evaluation shall be done with the help of a technician because the characteristics of the process are frequently involved. Nevertheless, in some cases the answer is contained in the description of the different factors. The point is to try to understand what happen to the effects of a factor that undergo an average.

As said in section 9.3, the information of bias can be of two types: the value and uncertainty of the systematic error to be corrected or the indication of an interval in which the amount of the systematic error can fall. In the first case the uncertainty given is connected with the random effect observed in measurement replications when the calibration was made. It is certainly reasonable to think that similar effects happen in the working phase, that is, the results taken with our replicated measurement can feel the same factors and can produce effects of the same range of the effects observed during the calibration.

In our case the situation seems different, because we do not have a declared value of the systematic error, but only an interval of variability. The amplitude of this interval would show clearly to a technician that it is not the residual uncertainty after the correction of the systematic, but the total interval where the systematic, of different instruments of that type, can fall. But the systematic is typically constant for a given instrument, therefore for our instrument it will not change in our replicated measurements. The effects on our replicated measurement will be confined in a part of the interval of variability we have indicated, therefore their average is not expected to go near the reference value, and formula 10.17 is not working.

We have completed the data for bias, one of the most complicated of the table as we will see. We pass now to the second factor: the resolution.

Here the situation is very simple: it is a Type B factor non-oscillating (as we have seen in section 3.2.1, resolution is constant or, for multirange instruments, changes by steps).

The value of resolution is the total variability interval; therefore it shall be divided by 2 to get the half-interval to insert in column 9 and, since it is not oscillating, we put the coefficient 3 in column 10. As there are no doubts on the resolution (it is sufficient to have a look to the indication scale), D.o.F. are 100.

The resolution applies to individual data, therefore no problem of input multiplicity. For output multiplicity the variability interval declared is 1 reading interval, while for technical instruments the variability is mostly ± 2 intervals; therefore the 5 replicated measurements can arrive in any place of the declared variability interval (and more), so that equation 10.17 can be applied and $n_r = 5$.

The next factor is the precision. As said in section 3.2.4, it can be expressed as repeatability if it is the effect of the measuring instrument alone, or as reproducibility if it is the total effect of instrument, operators, environmental conditions and measurand. In our case we have both data, and we use reproducibility, as it is comprehensive of all effects. Reproducibility was evaluated, as described:

- reproducibility was tested by 10 replications made by all operators, during all working periods, obtaining a variance of 10.7 μm^2;

We get the value of variance, which is a statistical information, but in our table we do not have a column for the variance.

The reason is that we have column 8 for standard deviation; therefore we can calculate the standard deviation as square root of the variance to be inserted in the column 8. As it is declared that the variance was obtained using 10 data, the D.o.F. are 9. We do not have a statement that the variance is that of the average, therefore it refers to individual data and input multiplicity $n_d = 1$.

For output multiplicity we shall really pay attention to the way reproducibility was evaluated and the 5 replicated measurements were done.

About reproducibility, as it has been clearly said, the whole measurement complex, instrument, operators, environmental conditions and measurand were involved. Certainly this is not the testing condition; the piece to be measured is taken by one of the operators, at a given time of the day, and in few seconds the five measurements are done. Therefore, the five measurements do not sense the effects of the different operators and of the environmental conditions; in this case, the variability of the repeated measurements is smaller than the variability declared for reproducibility and equation 10.17 does not work: consequently $n_r = 1$.

Next step is the variable α, the thermal deformation coefficient, whose value is taken from a handbook.

Variability is not declared in the handbook, so, as trial, two units of the least significant digit are taken as Type B variability. For the D.o.F., we have no indications on how believable is the handbook used (but in the real work we can judge the authoritativeness of handbook used), therefore we can choose 30 as the middle range situation.

No multiplicities involved (variability take into account mainly the possible difference of the real material of the workpiece from the nominal material defined in the handbook). The variability guessed of 2 units of the least significant digit does not refer to any average, so $n_d = 1$, and the replicated measurements do not produce variations of the thermal deformation coefficient; therefore it is not possible to see the effect of this factor in replicated measurements, consequently $n_r = 1$.

Eventually we arrive to the temperature. If the aim of our analysis is to give an indication of the possible difference of results on a workpiece that can be measured by any one of the operator, at each hour of the day or the night, that is, in each possible temperature condition, the whole temperature interval over 24 hours shall be considered. A measurement done at any time will certainly sense the day-night sinusoidal pattern of the temperature; therefore the half-interval of 5 °C is put in column 9 and coefficient 2 (oscillating factor) in column 10. The temperature effect act on individual measurement, therefore no input multiplicity and replicated measurement, as said before, are done in a short time; in this case it will not sense the oscillating pattern of temperature, so no output multiplicity.

The table calculates automatically the variances $u^2(x_j)$ of each factor using the square of the standard deviation of column 8 or the relevant formula for Type B contributions.

10.2.3 Variance propagation

The previous steps allowed to evaluate the variances and the equivalent variances of the independent variables $u^2(x_j)$. Now we have to propagate them to $u_j^2(y)$, the contribution of each factor to the variance of the dependent variable. We shall recall the condition of use of the simplified formula 10.16, which is valid if, inside of the variability intervals, the conditions of linearity and independence between the independent variables held. For linearity there is no problem, but in very strange cases. For the independence it can happen, even if not frequently, to have interactions. In case the technical part suspects interactions, it is necessary to evaluate the variance-covariance matrix and to use the complete formula 10.13. If there is no reason to suspect interactions, the simplified way is adopted.

What is required to do the calculations is to evaluate the partial derivatives. This can be done in the analytical way if the mathematical model is not complex; in our case we have:

$$L_{20} = L_T[1+\alpha(20-T)] = \mathrm{G}(L_T, \alpha, T)$$

$$\frac{\partial \mathrm{G}}{\partial L_T} = 1+\alpha(20-T)$$

$$\frac{\partial \mathrm{G}}{\partial \alpha} = L_T(20-T)$$

$$\frac{\partial \mathrm{G}}{\partial T} = L_T\alpha$$

If the mathematical model is complex, it is easy to resort to numerical evaluations of the derivatives. In any case we arrive to get the values of the sensitivity coefficients c_j and to complete the evaluations of the contributions of each factor to the variance of the dependent variable. Note that the derivatives, obtained by a mathematical operation, are defined for the variables, not for the factors; therefore, for instance, the derivative is calculate for the variable L_T and used for all its three factors: bias, resolution and precision.

At this point, having all values of $u_j^2(y)$, we can complete the application of PUMa method. First we can check the importance of the guessed factors. In our case the variability assigned to α was guessed to be 2 units of the least significant digit; now we can see that this produces a contribution of 3.4×10^{-12}, which is certainly negligible against the contribution of 9.7×10^{-10} due to the temperature; therefore no reason for revising the variability guessed for α.

11 THE EXPANDED UNCERTAINTY

11.1 EXPRESSION OF THE EXPANDED UNCERTAINTY

In the previous chapter we arrive to the definition of combined uncertainty, which is a standard uncertainty of the dependent variable, collecting the effects of all factors involved. Following the procedure described there, we can evaluate the variance $u^2(y)$ of the dependent variable Y, thereafter it is simple, doing the square root, evaluating the standard uncertainty $u(y)$, which is the combined uncertainty of the dependent variables. But the standard uncertainty gives an interval with a risk of error of more than 30%, too large. It is very useful for calculations, but not a good indication of the variability of Y, as there is more than 30% probability to get values outside.

As stated in sections 5.5.7 and 5.5.11, it is necessary to consider an acceptable risk of error of the first kind and to evaluate the corresponding confidence interval, which is called by GUM *Expanded Uncertainty*, or, in the general use simply Uncertainty. While the standard uncertainty is indicated by the symbol u, expanded uncertainty is indicated by the symbol U (upper case). Sometimes, when the uncertainty is given in relative form, the symbol W is used. The procedure is simple as the calculation of the confidence interval, but to have a wider generality we shall use Student distribution and Student t. Given the risk of error α and the D.o.F. ν, it is possible to evaluate t. Again GUM uses different symbols and terminology from the classical Statistics, therefore Student t assumes the name of coverage factor and the symbol k. The confidence half-interval, given by $t \cdot s$, became the Expanded Uncertainty taking, with the GUM symbols, the form:

$$U = k \cdot u \tag{11.1}$$

So that the uncertainty is frequently declared, together with the measured value m_y, in this form:

$$y = m_y \pm U \tag{11.2}$$

To calculate the coverage factor, which is a Student t as said before, we need the values of α and ν. The risk of error is determined considering the relevant consequences, but the number ν_y of degrees of freedom of $u(y)$ shall be calculated from the data of independent variables.

11.2 DEGREES OF FREEDOM

The evaluation of degrees of freedom (D.o.F.) is not hard on directly measured factors. Their number is calculated from the data number minus the number of constraints. If for evaluating the standard deviation of n data the average, used in the standard deviation formula, has been calculated, a D.o.F. has been lost and the standard deviation remains with $\nu=n-1$ D.o.F. In the case in which a regression has been assessed with a mathematical model that contains n_p parameters, if they are calculated through the n experimental data, n_p D.o.F. have been lost and it will only remain $\nu = n - n_p$ D.o.F.

Also on the contributions of Type B uncertainty there are no major problems. In fact, one of the advantages of these contributions is being very "robust" and this data can be managed as described in section 10.2.2.

It is more difficult to evaluate the D.o.F. corresponding to a combined standard uncertainty due to several factors. In that case, in fact, we must consider how the individual factors, of which are known the D.o.F. ν_j, contribute to the D.o.F. ν_y of the variance of the dependent variable. To make this evaluation we use the Welch-Satterthwaite formula that takes into account the D.o.F. ν_j, weighing them in respect to the variance contributions $u_i^2(y)$ of each independent variable X_j, and the variance $u^2(y)$ of the dependent variable, relating them with the D.o.F. of the dependent variable ν_y:

$$\frac{u^4(y)}{\nu_y} = \sum \frac{u_j^4(y)}{\nu_j} \tag{11.3}$$

From which is easy to calculate the unknown D.o.F. ν_y:

$$\nu_y = \frac{u^4(y)}{\sum \frac{u_j^4(y)}{\nu_j}} \tag{11.4}$$

This calculation can be organized how shown in Table 11.1.

We can apply this with our example of length measurement as sown in table 11.2. We can check the effect of the 5 guessed D.o.F. for bias, comparing data in the last column, and we can see that it is not an important factor, giving a contributions of $7 \cdot 10^{-24}$ against $3.1 \cdot 10^{-20}$ due to temperature. To complete it, we consider the risk of error α, but, as we do not know the consequences of a false indication of the confidence interval, we use the conventional value of 5% so that the confidence level is 95%, and now we can calculate Student t and evaluate the expanded uncertainty U.

This table is very useful. It is a widely accepted way for expressing uncertainty; it allows to fulfil the requirements of Quality Standards, but, more important, give us the possibility of using correctly the measurement results, as it was shown in Chapter 1.

But the main advantages are due to the support it can give when one shall manage non-exact variables. This happens nearly always not only in measurement (instrument, operator, ambient and measurand parameters) and production (pieces and process parameters), but also design is affected by data that are not completely known (material characteristics, real shape and dimension of the workpiece, forces involved). Many problems of production processes can be fixed. On the technical side the problem is frequently to get a product more close to the requirements; a table on the uncertainty of the qualifying characteristics of a product can highlight which are the main factors producing variability, by comparing the contributions of each factor to the variance, so that the correcting action can be effective and limited to the factors really important. On the economical side the problem is to reduce costs and at the same time to keep the quality level. Again the comparison of the contributions to uncertainty allow to discover the negligible factors, which can become worse without affecting the quality of the product, that is, on which is possible to save money.

Tab. 11.1 Part of the uncertainty table regarding the calculation of the D.o.F. ν_y

x_j	s_j	a_j	ν_j			$u_j^2(y)$	$\frac{u_j^4(y)}{\nu_j}$
X_1							
...						...	...
X_q							
Variance $u^2(y)$						$\sum u_j^2(y)$	$\sum \frac{u_j^4(y)}{\nu_j}$
Standard deviation $u(y)$						$\sqrt{\sum u_j^2(y)}$	$\sqrt{\sum \frac{u_j^4(y)}{\nu_j}}$
Degrees of freedom ν_y						$\nu_y = \frac{u^4(y)}{\sum \frac{u_i^4(y)}{\nu_i}}$	
Confidence level						P	
Coverage factor						$k = t_{\alpha;\nu_y}$	
Expanded uncertainty						$U = k\,u(y)$	

A third very important advantage is that we can plan our process. For instance, for a measurement or a testing process we must decide which class of instrument shall be used, how good shall be the environment conditions and how many measurement to be averaged shall be done to reduce the effect of random errors: the uncertainty table can be prepared using type B contributions without the need to have measurement results. This allow us to try, to check the expected uncertainty and, if necessary, to modify the measurement scheme.

11.3 Examples of expanded uncertainty calculation

Some examples will help to better establish the above cited concept.

A first example can be the calculation of the expanded uncertainty on the basis of a series of experimental data. This is very common, but can occur in two different scenarios. The first case is the one corresponding to the decision to reduce the effect of the random errors, and this is achieved only replicating the measurements ***n*** times and taking, as information, the average ***m*** of the obtained measurements.

Tab. 11.2 Table of the uncertainty budget for the example of length measurement

Mathematical model $L_{20} = L_T[1 + \alpha(20 - T)]$

Variable x_j			Statistical					Non Statistical		Parameters estimated						
Symbol	***Value***	***Note***	U_j	P_{dj}	v_{dj}	k_{dj}	s_j	a_j	k_a	v_j	n_d	n_r	$u^2(x_j)$	$c_j=\Delta y/\Delta x_j$	$u_j^2(y)$	$u_j^4(y)/v_j$
L_T	0.800074	Bias	3.7E-06	99.7%	100	3.0	1.2E-06		3	5	4	1	5.9E-12	1.0E+00	5.9E-12	7.0E-24
		Res.		95.0%	100	2.0	0.0E+00	5.0E-07	3	100	1	5	1.7E-14	1.0E+00	1.7E-14	2.8E-30
		Prec.		95.0%	100	2.0	3.3E-03		3	9	1	1	1.1E-11	1.0E+00	1.1E-11	1.3E-23
α	1.1E-05	Table		95.0%	100	2.0	0.0E+00	2.0E-06	3	30	1	1	1.3E-12	-1.6E+00	3.4E-12	3.9E-25
T	22	Estim.		95.0%	100	2.0	0.0E+00	5.0E+00	2	30	1	1	1.3E+01	-8.8E-06	9.7E-10	3.1E-20
y=L_{20}	800.056	mm														

Variance of y, $u^2(y)$	9.9E-10	3.1E-20
Standard uncertainty of y, $u(y)$	3.1E-05	
Degrees of freedom of y, $\nu(y)$	31	
Confidence level	95%	
Coverage factor (Student ***t***)	2.0E+00	
Expanded uncertainty $U(y)$	6.4E-05	m

The second case is related to the qualification of a measuring complex, which is an analysis of detailed reproducibility used for a robust calculation of the relevant uncertainty produced in the event of later use; for instance Quality control of products, where the measurement procedure will provide for a single or a reduced number of measurements.

11.3.1 Reduction of random errors effects through repeated measurements

We have to calculate the uncertainty to be attributed to the average $\boldsymbol{m}$ of $\boldsymbol{n}$ measurements. To do this, we need to evaluate the standard deviation of the average distribution, but from the available data (the $\boldsymbol{n}$ results x_i) we can only calculate the standard deviation $\boldsymbol{s}$ of the individual data.

We can, however, estimate the standard deviation of the average with the formula 5.17. Such method is also convenient when you have a sufficient number of data divided into q groups of $\boldsymbol{n}$ data.

It is possible to calculate the q averages and the relative variance s_m^2, but this way the variance would only have $\nu = q-1$ D.o.F., while calculated with the formula 5.17 would have $\nu = q \cdot n-1$ D.o.F. that are many more.

In practice, we operate how follows, using data of table 11.3:

Tab. 11.3 Sample data for evaluating the reproducibility and reducing the uncertainty due to random effects

Experimental data [mm]		
30,05	30,07	30,13
29,99	30,06	29,92

Required level of confidence $\mathcal{P} = 0{,}97$

We calculate the standard deviation $\boldsymbol{s}$ from $\boldsymbol{n}$ experimental data;

$\boldsymbol{n} = 6$

$\boldsymbol{s} = 0.07$ mm

if we take as information of the measurement outcome the average of $\boldsymbol{n}$ experimental data, we will have to estimate the uncertainty of the average using the standard deviation s_m of the average distribution. This is calculated using the formula 5.17.

We calculate the coverage factor $\boldsymbol{k}$ as Student's $\boldsymbol{t}$ corresponding to the confidence level $\mathcal{P}$ required. For this purpose, we need to know the number of D.o.F:

$$\nu = n - 1.$$

Number of data $n=6$; number of D.o.F. $\nu=5$

Required confidence level $\mathcal{P} = 0.97$

$$k = t_{P;\,\nu} = t_{0{,}97;\,5} = 3.00$$

we calculate the expanded uncertainty $U_m=k\cdot u_m=k\cdot s_m$

$$U_m=k\cdot u_m=3.00\cdot 0.03 = 0.09\ mm$$

In case the variability of our data is random, we can estimate the expected uncertainty in case we do the average of a different number of data, just doing the same calculation and changing the value of n.

For instance, we can check that the expected uncertainty, if we do the average of 12 data, will decrease to 0.06 mm.

11.3.2 Qualification of a measurement's complex through repeated measurement

The context of management of a production control measurement complex, represents a case of multiple data management. It is not possible, in this case, to perform numerous measurements of each workpiece to be controlled because of their high number. Remains, therefore, the problem of defining the standard uncertainty of the measurement complex.

This should be done, whenever possible, evaluating the effects of the involved influence factors as Type B contributions.

Often, however, the measurement complex has operating conditions specific to each individual case. It is therefore appropriate, in the qualification phase of the measurement complex, to do a test of reproducibility replicating several times the measurement of a single workpiece representative of those that will be controlled.

It is much better spending time in the qualification of the measurement complex, adequately determining its contribution to the uncertainty, rather than repeatedly replicating the measurements on each workpiece. In the latter case, in fact, the number of D.o.F. of the standard deviation would be low, thus penalizing the estimated uncertainty (increase in the Student's t). Moreover, the total amount of measurements required would be very high.

In practice it is convenient to perform a series of preliminary measurements, representative of the conditions of use of the measurement complex (reproducibility test) and determine from these measurements the value of the standard deviation characteristic of the measurement's complex.

In the preliminary stage the n_1 measurements shall only determine in a good way the reproducibility(i.e. the standard deviation), therefore n_1 shall be as high as possible; the average in this phase is not of interest.

The aim is to evaluate the expected uncertainty for the successive routine control measurements, which will be done with a single measurement or with a limited number n_2 of measurements. The standard deviation to be considered in this phase is the one of individual data s. If the control procedure provides n_2 measurements with a standard deviation s_2, and the result of the routine control will be the average of these measurements, we should cumulate the knowledge on the standard deviation doing the average of the variances s^2 and s_2^2 weighted on n_1 and n_2 as individual variance (attention! Standard deviations cannot be added; therefore the average shall be done on variances), than take into account the multiplicity n_2 and evaluate the standard deviation accordingly.

Let's examine some experimental data, given in tables 11.4 and 11.5:

Tab. 11.4 Qualification data for a robust evaluation of reproducibility

Qualification data [mm]							
29,96	29,94	30,04	29,98	30,08	30,03	30,03	30,03
30,00	30,04	30,00	30,01	30,07	29,96	30,07	30,09
29,93	30,06	29,97	29,92	29,94	29,97	29,92	29,99
29,96	29,96	30,01	29,97	29,92	29,99	30,01	30,01
29,94	29,98	29,98	30,05	29,98	30,07	29,99	30,01

Tab. 11.5 Repeated data for reducing the uncertainty due to random effects

Routine data [mm]		
30,05	29,99	30,08

We calculate the standard deviation s from n_1=40 qualification data and s_2 from n_2=3 routine control data:

$$s = 0.0473 \qquad s_2 = 0.0458$$

The expected value of the standard deviation of individual data and the relevant D.o.F. are given by:

$$s_{ind} = \sqrt{\frac{n_1 s^2 + n_2 s_2^2}{n_1 + n_2}} = 0.0472 \qquad \nu = (n_1 - 1) + (n_2 - 1) = 41$$

Thereafter we consider the multiplicity n_2 for evaluating the standard deviation of the average of 3 data and the relevant Student t (confidence level 97% and 41 D.o.F.):

$$s_m = \frac{s_{ind}}{\sqrt{n_2}} = 0.0272 \qquad t = 2.25$$

that allow eventually to evaluate the expected expanded uncertainty $U = 0.06$ mm. The final result y is given by the average m of routine measurements with the relevant uncertainty:

$$y = m \pm U = 30.04 \pm 0.06$$

Note the attention on the number of D.o.F., calculated considering the number of data used for evaluating each standard deviation, and the multiplicity, calculated on the number of data averaged.

11.3.3 Planning the number of repetitions

The analyses described in the previous sections are made a posteriori. This is usual, when data are available, or used as shown before in a qualification test for determining the initial information, when there is no other source. On the other side, the routine activity shall be planned mainly on the number of data necessary for reaching the prescribed tolerance. As shown in section 1.2.2, ISO 14253-1, there is a relationship between tolerance and uncertainty; as a rule of thumb, one can say that uncertainty shall be lower than 1/5 of the tolerance. Therefore, if for the routine test given before we want to verify a tolerance of 0.2 mm, the procedure adopted is not adequate: we shall have an uncertainty lower than 0.04 mm. We shall plan the number of routine tests to get this goal. It is easy to go by trials. Testing with n_2 = 7 we get the required expected uncertainty of 0.04 mm.

11.4 USE OF THE EXPANDED UNCERTAINTY'S CALCULATION TABLE

The case of length measurement, organized with the scheme of Table 11.2, can show how to use the uncertainty table. If the expected uncertainty is not acceptable, we have to define the corrective actions to reach the required value. If you let this charge to the technical part, probably they will prepare a list of all possible factors (even impossible), but usually they are not aware on which are the main factors to be tested before.

Going back to table 11.2, we can see from the values $u_j^2(y)$ that the important factors are the temperature variation and the reproducibility on L_T measurements. The question for the temperature is that a guessed variation covering day and night was used: the use of a thermometer will be of good help. At least two ways are possible: a measurement each hour is really practicable, with a guessed variation in that time interval of less than ± 1 °C, or a more accurate temperature measurement associated to each length measurement, letting possible differences of about ± 0.5 °C, but nearly doubling the cost of the test. With the uncertainty table we can see immediately the effect of these hypotheses, just changing the guessed variation of temperature and the coefficient k_a, which, being the variability no more oscillating for the short time interval, become 3. The expanded uncertainty on L_{20}, from $6.4 \cdot 10^{-5}$, becomes respectively $1.4 \cdot 10^{-5}$ and $1.0 \cdot 10^{-5}$, allowing to choose an acceptable way. On the other side, the resolution of the instrument used for measuring L_T is unnecessarily good; therefore, with the aim of reducing costs, it is possible to evaluate if a less performing instrument can be use just in the same way, replacing the characteristic of the proposed instrument in the uncertainty table.

Made in United States
Orlando, FL
04 November 2026

90206775R00138